VIRUSES

AGENTS OF CHANGE

ALSO BY THE AUTHOR
The Truth about AIDS: Evolution of an Epidemic
Potpourri, Incense and Other Fragrant Concoctions

VIRUSES

AGENTS OF CHANGE

Ann Giudici Fettner

McGRAW-HILL PUBLISHING COMPANY
New York   St. Louis   San Francisco   Bogotá
Hamburg   Madrid   Mexico   Milan
Montreal   Paris   São Paulo   Tokyo   Toronto

The book is not intended to replace the services of a physician, nor is any of the information presented herein intended as medical recommendation and/or advice in any way whatsoever. The reader should consult with his or her own physician concerning his or her own health needs.

1 2 3 4 5 6 7 8 9 DOC DOC 9 5 4 3 2 1 0

ISBN 0-07-020664-3

Library of Congress Cataloging-in-Publication Data

Fettner, Ann Giudici.
    Viruses : agents of change / Ann Giudici Fettner.
      p. cm.
    ISBN 0-07-020664-3
    1. Virology—Popular works. I. Title.
    [DNLM: 1. Viruses. QW 160 F421v]
QR364.F48 1990
616'.0194—dc20
DNLM/DLC
for Library of Congress   90-5628
                                                      CIP

Book design by Eve Kirch

For my children and their children and with special thanks to my brother-in-law, Dr. Robert C. Cauthorn. But mostly in memory of Mark Bell and the multitude of other friends lost to AIDS.

Contents

PART TWO: THE HOST

PART THREE: THE DISEASES

Foreword

by Larry Kramer

Every era has its great tragedies. Our era's greatest tragedy is AIDS, a putrid virus that is evidently going to mow down huge swaths of the world's population before world leaders decide to *really* try to stop it.

If we're lucky, great tragedies have heroes. But little about AIDS has been lucky. Many of its greatest heroes, like AIDS itself, have gone unrecognized, or been ostracized, scorned, and shut out of the "mainstream" by the second-raters who always seem to wind up calling the shots in governments that are more and more bureaucratic, and a world that cares less and less.

I often wonder whether those who *have* the valuable insights, those who are gifted with the qualities most necessary in time of tragedy, are also those who have "tragic flaws." Are they, by temperament, simply not team players? Are they unable to cope with those who are? Are they too temperamental? Too opinionated? Too docile? Not diplomatic or political or practical enough? Are they *too* smart? Or are these tragic flaws not theirs, but society's? After all, why should good minds go to waste? How can anyone be thought *too* smart?

This book is full of tragic instances in which those who turned out to be right were, initially and for too long, ostracized and condemned by those who turned out to be wrong, thus allowing endless horrors to proceed unabated. It is a book written by someone I consider a hero. She's someone who most emphatically belongs to that small group of superbly gifted people who possess insight and knowledge and the ability, not only to see what is happening, but to interpret it in exceptionally useful, indeed visionary, ways. She is someone who, like the rest of her breed, has been tragically shut out by the "mainstream" from rightful participation.

Right from day one of the AIDS pandemic, Ann Giudici Fettner should have been writing about the virus for *The New York Times* or *Time* or *Newsweek* or some similar "journal of record" that reaches millions. Had that happened, the world would have known more about AIDS and understood it more completely and more swiftly than it has. She is among a very few who knew exactly what was happening—and, who foresaw what did happen and what should have been done, early, before it was too late. From the outset, she had an acute sense of what should have been researched, which doctors and scientists should have been called in, what tasks should have been assigned and performed by which agencies of national and world governments.

But fate, with its own tragic flaws, dictated otherwise, and it fell to *The New York Times*—"all the news that's fit to print"—to utilize the services, one after another, of a series of quite possibly the worst AIDS reporters in America. And it has been just about the same with other publications. So, day after day, week after week, month after month, year after year, the world has simply not known, understood, or comprehended what was going on about AIDS. Or been allowed to hear any original ideas about how to cope with it or cure it.

I know that this is an amazing suggestion I am making: that a writer might actually have accomplished what an army of medical men and women have not—that if Ann Fettner had been there, warning, predicting, cajoling, extolling, suggesting, putting two and two together in her incomparable way, AIDS might not have grown so awful.

But that's how much I respect Ann Fettner's brains and insight and historical perspective. That's how much I respect her hunches. Like a good hostess who puts the perfect guest list together—strangers who, when seated next to each other, are suddenly capable of enthusiasm, insight, and wit that they never evidence alone—Ann knows how to maneuver graciously and effectively through the tricky shoals of science and scientists, doctors, and disease. She would have said that she thought Dr. X should talk about this with Dr. Y. And Dr. X and Dr. Y would have talked together, even if they had hated each other for decades, because Ann would have been with them. Together, they would have brainstormed and posited and come up with something or other that would have shaved a few months off here and there. And Ann would have repeated this procedure, all along the way, with Dr. A and Dr. B, and Drs. C, D, E, and F. She would have written about all this in a regular column in *The New York Times* and it would have been read all over the world and inspired further thinking, perhaps in some foreign laboratory or American university.

But Ann didn't write for the *Times*. The *Times* would never hire anyone as prescient and able and knowledgeable and in love with science as Ann. And neither would anyone else in the mainstream press. Because to work for the mainstream press you have to suppress your bubbling excitements and enthusiasms and revolutionary hunches. You have to be a team player, and most of the really smart, exciting, innovative, creative, essential minds in this world—as this compendium of historical virology Ann has written makes abundantly clear—are not team players. And the tragic failure is not only the boring *New York Times*'s and the wretched second-rate media's, but ours—we who are suffering because the knowledge that these essential but excommunicated minds bring to any tragedy is (I hope this is obvious) the kind of knowledge that can end the tragedy.

You are about to read an enthralling book. Ann's knowledge pours out in a furious, rapid, tumbling excitement that only someone in love with science can convey. She makes you feel the *disorganization* of it all, the messiness, not only of science but of the universe, with all its bold steps forward and its blind mistakes. She sees the scornful stupidities and the tiny nuggets of celebration. She stews everything together—rightly, because, like everything else, science, medicine, and viruses are messy.

Yet somehow this messiness is part of the enormous puzzle that is pieced together, little by little, century by century, laboratory by laboratory, scientist by scientist, mistake by mistake, somehow. . . .

Her love for what she writes about is felt on every page.

What does it mean to be in love with science?

I think it means that you want people to live. You want to help people learn, because knowledge helps you to live, longer and better. You want to give to them the supreme gift: a better, healthier, happier life.

And this love of science is filled with obsession, because you are driven by the utmost urgency, because each day is a precious day, and there aren't all that many of them in our allotted givens. And you know, not only how fast life flows, but how long it takes to get anything done. You know how turgid, unyielding, and ungenerous the entire system of research funding has become, and how hard, if not impossible, it is for bright young minds to find official succor or patronage. You know the way fine minds turn to other, more remunerative, fields because it has happened to you—because there's a family to feed, and because our government (and our culture) does not prize, as other governments (and cultures) most certainly do, the courageous, inventive, inquiring intelligences that oh so desperately want to delve into the mysteries of science. You know that this is why we fall behind so much each year, that this is why discoveries are made by scientists in other countries, spending a fraction of the money that is thrown away in our wasteful land. You know that bright young minds need not only money and patronage, but a modicum of the freedom that created our country but that has since disappeared, as president after president cuts budget after budget, hoping that assembly-line science and the regimentation of clonedom can still, nevertheless and somehow, bring results. You know that these results do not come, or if they come, they come too late to save a son or a daughter or a baby or a hundred of them, or ten thousand of them, or some millions of them—each one somebody's son or daughter, who, had only the second-rate not been in charge of the store, might somehow. . . .

I guess you can tell that I think Ann Fettner is one classy person. I hope you like her book. And learn from it and from her. As I have.

Introduction

"Our preoccupation with acquired human immunodeficiency syndrome should not obscure the multiplicity of infectious diseases that threaten our future. It is none too soon to start a systematic watch for new viruses before they become so irrevocably lodged," writes Nobel laureate Joshua Lederberg.

In this matter-of-fact fashion, the scientist reminds us that AIDS is not the first, and by no means the last, novel viral disease we will experience. Many millions of new viruses must have surfaced and disappeared throughout the history of life on Earth, leaving evidence of their presence in folklore and literature and in the remains of ancient bones.

Bacteria appeared only a billion years after our planet was formed, and bacteria are complex entities. There had to have been an ancestor molecule that gave rise to such an elaborate creature with its genetic information hidden behind cell walls. What better to set the stage than the snippets of genetic material that are viruses?

Viruses appear to be as primitive a form of "life" as is currently known. By looking at the panoply of these submicroscopic entities and their profound effects on living creatures, we begin to understand what causes many mysterious illnesses and to start chipping away at the mystery of what constitutes the processes of life.

Even twenty years ago the study of viruses was mostly a matter of studying the diseases they caused. Today our interest has been stimulated by technological advances: We can see these pathogens for the first time, can walk their strings of molecules, can empty and refill their envelopes with toxic chemicals targeted against cancerous cells. In terms of treatment, we're only marginally better able to deal with viruses than we were 100 years ago, when they were known as filterable agents capable of causing the diseases that couldn't be blamed on bacteria, fungi, and other "giant" parasites.

Unlike those more familiar germs, viruses tamper with our genetic programming as they pass among us during sex and sneezes. Most do no damage; others, lurking unnoticed in cells, can generate cancers or change the architecture of our brains years later. Others, the ordinary viruses we thought we understood, are suspected of kicking off the processes that lead to arthritis, mental disorders, and lost pregnancies.

We are, in fact, born surrounded and inundated by viruses, extraordinary entities that are unlike what we commonly understand as life and that become activated only in the cells they "infect." Fortunately, few viruses pose a threat to humans, and most (but not all) that cause disease in animals and plants are of only secondary importance to us.

However, in our society, with plagues and dangerous epidemics seemingly under control, it has been easy to cling to the comfortable idea that an antibiotic or vaccination against childhood diseases can take care of serious infections. When we cover our faces to sneeze in public and wash our hands before eating, we tend to think we are not only protecting ourselves but are acting responsibly.

Then along came AIDS to remind us how powerless we really are despite our prevailing notion that humanity is superior to the rest of the natural world. Lederberg says that this notion fails to take into account the fact that there's no guarantee that we're inherently entitled to inherit the Earth. Looking back at the evolution of the animal world, in which gradual mutations of species once seemed to follow a logical order, we are generally unaware that viruses can make genetic shifts, can turn into completely new entities, with lightning speed.

So far, no pathogen has totally wiped out life, although, given the nature of viruses, one conceivably could do so and hang around in its strange unalive state until other creatures evolved to supply its needs. But unlike humans, who for the most part seem to assume that they can destroy the oceans, the rain forests, and the atmosphere without killing our Earth host, viruses adapt to the reality of their situation. If it's a matter of survival through accommodation, the smart money has to be on the viruses.

AIDS caught us unprepared. It appeared with all the trappings needed to attract attention, beginning by killing "exotic" populations. It was linked to socially forbidden activities and traced back to the "dark" continent of Africa, about which few Western people have much knowledge. For several years early in the pandemic there was only speculation about its cause, and modern civilization faded into the background because of the ancient fear that something mysterious and evil had been let loose among us.

The uninformed and the vicious called for quarantining the infected, and the miasma of guilt and blame that accompanied the first few hundred deaths lingers still. Rumors of conspiracy, of an accident in germ warfare development, even of the deliberate release of a manufactured pathogen into Third World populations, continue to circulate. They are taken seriously by much of the public, to whom it is inconceivable that such a lethal event could have taken place without the hand of humankind.

We take too much credit and ignore what is going on around us every day, forgetting that new influenzas surface regularly, that chickenpox can resurface fifty years after a childhood infection, that madness can be caused by measles, and that cancers follow some infections. Except for smallpox, the great diseases of contemporary humankind still sweep through underdeveloped nations: malaria, tuberculosis, cholera, polio, yellow and dengue fevers, measles, mumps, and chickenpox, plus a never-ending series of enteric parasites that blind, weaken, and kill millions of people evey year. Now seals are dying from dog viruses, dogs from a pathogen once restricted to cats and mink, and humans and monkeys from AIDS.

When AIDS came to public attention in the early 1980s, and particularly as rumors of a human-made and deliberately released agent began to circulate, I was reminded of biologist Lyall Watson's account of the change 250 years ago in the property of glycerine, a sweet fat used in medicine and explosives. Despite every attempt to make it crystallize, glycerine remained liquid. Then "something strange happened to a barrel of glycerine in transit between the factory in Vienna and the regular client in London," writes Watson.* It arrived at its destination in a great crystalline lump.

When it was portioned out in small samples to delighted chemists, something even stranger occurred: In the presence of the glycerine crystals sealed in airtight containers, the liquid glycerine in labs began to crystallize, too. Watson says offhandedly that this "is a regular occurrence in organic chemistry. Yesterday something was impossible and today it is easy—partly because of the introduction of a new technique, but also in part because of the existence of a new state of mind."

The same elements have come together to create the AIDS pandemic. The technology is intercontinental air travel, and the state of mind is the sexual freedom that has become the norm in cities during the past twenty-five years. As frightening as the idea of this deadly new virus is, its appearance is merely a reflection of the massive social and technological changes that have taken place during this century. Without these changes, the human immunodeficiency virus might never have come into being or the virus would have remained hidden away on the fringes of a rain forest, killing only the unlucky few who came into contact with it. But just as the liquid glycerine required an unknown set of circumstances to form crystals, so the new virus was "shaken" into a novel configuration.

Most diseases fit within the analogy of crystallized glycerine. Something changes, and the rates of coronary heart disease, certain types of cancers, and mental disorders rise or fall. Searching for that something, we often point the finger of blame at environmental agents, foods, or, when it's frightening enough, various groups of people as having caused a particular illness.

It is the proposition of this book that the entities called viruses underlie many mysterious, widespread conditions, that these submicroscopic pathogens provide the "shaking" that disarranges the information exchanged by normally functioning cells and causes them to change their signals in ways that create illness. In most cases, we know as little about these changes as we do about the bizarre transformation of the glycerine, but we're learning fast. And AIDS, because it has arisen in tandem with at least partial knowledge of viruses and their effects on cells, is providing a rare opportunity

*Lyall Watson, *Lifetide* (New York: Bantam, 1980).

to examine both the scientific and social consequences of viruses in previously unimagined ways.

Great plagues have had far more dramatic effects on populations, but AIDS alone has spanned the history of human evolution, within decades moving from primitive African rain forests to the twentieth century. Lurking unnoticed in promiscuous urban populations and in the blood supply, the new pathogen enlarged its growing pool drop by drop until it reached a critical mass and exploded.

For the first four years after the identification in mid-1981 of an apparently novel disease syndrome, physicians, the press, and the public speculated about its causes. Finally, in the spring of 1985, Secretary of Health and Human Services Margaret Heckler, croaking with laryngitis, announced the identification of a new virus—a retrovirus—as the probable cause. The immunology researcher with whom I was sitting blanched and groaned.

"Oh, God," he said, "we're in big trouble."

Indeed we are, although at the time few understood the implications of AIDS being caused by that novel family of viruses.

But we're also going to be the beneficiaries of the exploding knowledge of viruses engendered by our having to deal with human immunodeficiency virus (HIV), the virus that starts the cascade of biological events that lead eventually to AIDS.

What image do you conjure up when the doctor says, "It's a virus"? Extraordinary electron micrographs show bacteria, fungi, or blood and cancer cells as sharp entities endowed with a purposeful, dynamic quality. But viruses reproduce photographically only as fuzzy outlines of proteins surrounding vague bars or dots of nucleic acids. Viruses are of a size that doesn't lend itself to portraiture, running a submicroscopic gamut of about 0.02 to 0.3 micrometer (a micrometer is one-millionth of a meter). This makes disagreements about them reminiscent of arguments about the number of angels that can dance on the head of a pin.

Imagining a virus takes the same leap of faith that trying to imagine infinity or "before the beginning" does. How is it possible for something to be not alive and not dead at the same time? Viruses are like packets of seeds which, despite the apparent absence of life in the hard, dusty bits, contain the genetic information to produce a row of tall hollyhock or a runty ground cover.

Seeds are merely a potential waiting for the medium needed to produce life, a process we don't question because it's so familiar. Though more mysterious because we can't see them, viruses are in many ways the same as seeds. As surprising as a crop produced by a jumbled-up collection of flower seeds would be, the "planting" of an unknown virus is just as puzzling. Is it dangerous? Unlikely: Most viruses do no harm, being longtime fellow travelers of humans, animals, and plants. But we are just scratching the surface of the known. The field is exploding with scientists whose determined pursuit of suspect pathogens, often coupled with fantastic luck, reads like a series of cliff-hangers.

But one doesn't have to be a scientist to know that viruses are mysterious. As esoteric and complex as science is, there's something logical underneath that is accessible to common sense. It is accessible, that is, when we let our imaginations wander and wonder about molecules as casually as we learned to accept the unknown X and Y of algebra.

Human immunodeficiency virus is relatively new as viruses go. Observing its present rate of mutation leads those studying it to estimate its existence to be no less than 25 years and no more than 100. Some biological change must have provided the catalyst to turn what might have been an insignificant African infection into a world-wide killer.

Watson writes, "Living together for mutual benefit, a process known as symbiosis, is the best possible way of producing dramatic evolutionary leaps. It involves no basic change in the genotype [genetic program] of either contributing partner, and yet it results in an association which takes both members into a whole range of new situations with new possibilities for survival and growth."*

Although we will never know when HIV mutated to infect humans or spilled over its original boundaries into the greater world, there's ample reason to believe it came, as life itself must have come, from the lush tropical rain forests of central Africa. Along the far reaches of western Zaire—a country four times the size of Texas but one with only the Congo River for transportation—lie Rwanda, Burundi, and Uganda.

In the mountainous corner shared by Uganda and Zaire are all the components of the inaccurate but popular idea of the enormous

*Lyall Watson, *Lifetide* (New York: Bantam, 1980).

continent: rain forests and heat, gorillas and other primates of many varieties, pygmies and small groups of the successful Bantu people. Largely isolated because of their inaccessible geographic situation and living in ancient ways, any of these people could have experienced what we call AIDS years before it was noticed in the west.

If this was the case, its spread was probably limited by tribal customs. But with the coming of colonialism, as the lure of a money economy and other spoils of civilization gradually became known in rural areas, people began working their way into rapidly growing urban areas to spread the virus to a wider population.

Because the epidemic appeared so suddenly, some people have speculated that the use of wild-caught monkey kidney cells to manufacture a vaccine—such as one for polio—might have unleashed the unrecognized and inadequately killed retrovirus virus that infects simians. Such a theory holds that the simian virus could have mutated enough to infect humans. A vaccination program intended to prevent one crippling disease could have created another: It's happened before. But the AIDS virus belongs to a family that is notorious for its quick-change artistry, so perhaps all it needed was a large enough pool of potential hosts.

Human immunodeficiency virus has drawn the attention of the world to viruses and seems to be unique and special. In some ways it is, but the range of problems this virus causes is merely an assembly of many of the effects known to be caused by other viruses. They are formidable: HIV comes with rashes, fevers, and diarrhea. It opens the way for the complicated family of herpesviruses and for fungal, parasitic, and bacterial infections. It wastes the body and the brain. It damages the blood and bone marrow. It allows the development of cancers and autoimmune disorders. Some of these conditions follow infection immediately; others take years to show up. Some infected people may not even experience significant ill effects during their ordinary life span.

Every one of these outcomes can be caused by other viruses— the usual ones we no longer think much about, such as measles, chickenpox, and influenza, and the esoteric agents that take half a lifetime to manifest. Even the simple phage virus that infects bacteria has special tricks hidden within its molecules. Viruses drive coronary artery disease, psychiatric illness, and autoimmune conditions. In the final chapter of this book, you'll come across something called a prion, which is known as an unconventional virus but in fact is a transmissible infectious particle, a kind of crystal thought to be responsible for causing dementia in the elderly.

AIDS and the virus that causes it figure prominently in this book. It is, after all, the current epidemic that has focused public attention on viruses, and this attention has brought to light the confusion most people have about viral infections. Ordinary words such as *infective, transmissible, contagious*, and even the old standby *antibiotic* are widely misunderstood and misused. It's no wonder fear has clouded the picture.

There's also been a lot of misapprehension about what scientists can and can't do. Much anger has been directed at the scientific community, which is suspected of dragging its heels in searching for a cure or vaccine. But the type of virus that causes AIDS has been recognized so recently that until this decade even the category in which it belongs was known only in regard to a couple of animal diseases. When you stop and think, you realize that all the doctor was ever able to do when you had a viral infection was tell you to rest and drink plenty of fluids.

This book is about a lot more than AIDS and the retroviral family to which it belongs. It's about the peculiar fact that a huge portion of the human genome consists of leftover bits of ancestral virus, some of which may lay the groundwork for cancers. It's about the viruses that might have caused a cousin's schizophrenia, a neighbor's diabetes or arthritis, or your child's hyperactive and rotten behavior when he or she has a fever blister.

Viruses infect everything alive. Honeybees are plagued by several that cause paralysis; mealybugs carry virus from one cocoa plant to another. Ants; turnips, radishes, and eggplants; oats, rice, and wheat; birds—all are infected by viruses. Who knows how many there really are? We've only begun looking for and at those that are medically or economically important to people, and we've done this with the new tools provided by the same technological revolution that has sent us AIDS.

The AIDS pandemic is moving science in totally new directions. Years of cancer research have produced little benefit, old diseases affecting millions have remained untreatable, and even the common cold is beyond our ability to manage. However, the urgent need to deal with AIDS is solidifying research that ten years ago was like liquid glycerine, sloshing around in the heads of a few researchers.

Because the AIDS virus allows ordinary pathogens to kill people, we've had to look more closely at them too. And because it effectively cancels out much of the immune system, a new kind of scrutiny has been turned on that vital network of cells. AIDS-generated knowledge is already being used against cancers and has produced a new class of antiviral drugs whose actions apply to other viral infections, such as hepatitis.

Answers to old questions are being explored in the light of AIDS: What keeps some of us healthy while others struggle with constant illness? What causes nongenetic birth defects, mental retardation, and learning disabilities? How can we make vaccines against unmanageable viruses? The desperate need to understand and deal with AIDS is illuminating the nature of the viruses that have plagued humankind throughout history and getting us ready to deal with others yet to come.

AIDS is a watershed in the study of human disease, and when you keep your antennae turned in the right direction, you can hear something happening just off stage as the shimmering crystals begin to coalesce and form new patterns in science and in society. Tomorrow another collection of molecules will drive the world of medicine. Today is the day of the virus.

PART ONE

THE AGENTS

1

Evolutionary Messengers

We can't start at the beginning because there isn't, and probably never will be, any evidence of the remote event that eventually gave rise to life on Earth. Without life, there still would have been change. The planet would have built and reshaped itself through the effects of water and shifting plates of rock, meteorites would have rained down, and volcanoes would have erupted. But change isn't life.

Life became a possibility when something began to reproduce itself. Some scientists think that something might have been crystal; others like the idea of mud, the primordial ooze. The trouble is that mere replication of self is just that—carbon copies. Such copies may lead to great success in quantity but not success of kind, not life. Somewhere along the way a wrench had to be thrown into the works that changed the "it-it-it-it" of replication to the "it-it-that-it-it-that" of reproduction. The "that" is what we are free to speculate about, and virus is a likely candidate, if not as a player in the initial event then surely for part of the rapid diffusion of the mixed bag of life-forms that exist on Earth.

All these life-forms evolved to the point where they could reproduce and survive, drifting up through the millennia with their functions and abilities virtually unchanged. The shark, spider, opossum, cat, and cayman; the kangaroo and pet dog all have refined their limited functions over time. In an evolutionary sense, they have no place left to go. This is true for fish and birds, reptiles and placentals, and the plants that ultimately sustain all life.

But human and virus seem to have evolved with a built-in mechanism for survival through mutation not merely of form but of function. Despite the grumbles of creationists, it's apparent that *Homo sapiens* departed from the ancestral line and left its close cousins swinging from trees or knuckle-walking across African plains, to become something entirely different. Charles Darwin's theories of evolution explained the biological diversity among Earth's creatures; today's exploration of what genes are and how they dictate what each living entity will be is beginning to reveal how natural selection can work at the molecular level.

As glorious or grotesque as the end results may be, human being, lobster, virus, and oak tree are similar assemblies of proteins whose production is directed by the few molecules that make up the genetic code of each one. As increasingly sophisticated methods are developed to look at the processes that eventuate in these different end results, their underlying structure and activities appear to be all the more similar

DNA and RNA

To reproduce itself effectively, a human, virus, plant, or single cell has to transmit a faithful copy of its genetic information to its offspring. It is also a part of the basic pattern of reproduction that some of the information is reshuffled before it is passed on. We know that to produce a child, the father and mother contribute their individual chromosome sets to the embryo. These sets mingle and re-form into a unique set which provides the offspring with his or her own novel genetic pattern.

Even *Escherichia coli*, the bacteria that populate the stomach, engage in a type of sexual behavior in which a pair of the bacteria establish a kind of bridge that allows them to swap some of their genetic information. In all creatures this information is contained in the two types of nucleic acid conglomerates we call DNA and RNA, which carry the messages of heredity.

DNA is made up of four simple chemicals—adenine, guanine,

cytosine, and thymine—that form the four-letter alphabet in which the information inherited by every individual is written. These basic chemicals are assembled in sequences that dictate the way in which they act. Genes thus are essentially short segments of DNA strung in a manner so critical that rearranging a single subunit may distort the design of a fetus or the ability of a virus to infect a host.

When the structure of DNA was discovered by Watson and Crick in 1953, the idea was that the chromosomes, the giant, double-stranded DNA molecules on which genetic information is encoded, contain the basic information needed to program every cell in an individual to function in a unique way. DNA was assumed to be the reservoir of all information, the orchestra leader whose baton determined the arrangement of the tune being played.

DNA was thought to pass its special information to its close cousin, passive RNA, which in turn simply follows the instructions of DNA and makes the proteins and other substances required by a living organism. But to accomplish all this, DNA needs a lot of help: special proteins and enzymes to act as catalysts, to splice and reassemble—in other words, to reproduce.

Viral Genes

It soon became obvious, however, that many molecular events cannot be accounted for by the DNA-to-RNA-to-protein model, particularly when viruses that maintain their genetic heritage in the form of RNA rather than DNA were discovered. As more information about these molecular occurrences has accumulated, it's been shown that some viral RNA is able to carry out a remarkable rearrangement of itself and to do this without the panoply of chemical helpers needed by DNA. To achieve this, however, viruses must first gain entrance to and infect a cell.

RNA appears to have unique attributes, some of which are shared by viruses and cells. But some viruses have the amazing ability to snip and splice themselves, leaving behind the genes that aren't necessary once they've infected a cell. Like cowherders who drive cattle into a pen and aren't needed anymore, the unneeded genetic material rides off into the sunset.

Imagine that the string of molecules that make up a virus's genome is like a two-foot ruler that reads left to right from one to twelve inches and then continues for another foot with the numbers diminishing to the one-inch mark on the right-hand end. Each inch mark represents a different gene, and some of the genes control activities that are essential for specific functions but are irrelevant to the virus's final job, which is to integrate its genetic material into the host cell, where it can take over the cell's works to reproduce.

If, for instance, the sections between the two-inch and four-inch marks at either end are needed only to penetrate the cell, once this function has been accomplished, those sections aren't needed any longer. Between the working parts of the genes are short strings of molecules called introns (a noncoding part of a gene) that serve to attach a gene to its neighbors.

If the central sections of the ruler are needed, for instance, to attach the virus to a cell's surface, they, along with the introns between them, are cut out of the ruler. Then the introns snip themselves out of the new assembly and turn back to the original ruler to splice together the two "raw" ends of the molecules left by their departure, thus creating a mature functional molecule. The free introns then splice themselves into a circle and presumably roll off and are lost.

Describing an experiment with introns, Roger Lewin wrote in *Science* magazine that "once the intron had excised itself from the precursor [the ruler], it nibbled off the first nineteen nucleotides" from one end of the RNA molecule. Intrigued by the ability of the introns to self-assemble without the presence of protein, Dr. Thomas Cech of the University of Colorado, whose investigations had proved that this amazing event happens, added a short RNA to the mixture in which all this was taking place. The intron promptly shortened some strings of molecules and lengthened others, setting about its business like a pattern maker with a pair of shears.

A genuine understanding of what this means is beyond the scope of anyone untrained in molecular biology. For our purposes, it's enough to know that an exotic chicken-or-egg argument was put to rest by this habit of the intron. Scientists had gone back and forth trying to determine how DNA could exist without proteins and how proteins could exist without DNA. Then Cech showed how an RNA molecule can react in and of itself without either DNA or protein being involved. As Lewin describes it in *Science* magazine, "The existence of RNA catalysts offers an intriguing glimpse of a former, more primitive age when the full range of metabolic and genetic machinery had yet to evolve."

In other words, it points back to a time when the water-locked planet began to dry, when free-swimming genetic strands would have needed a protective envelope and would have devised their own environments by enclosing themselves in membranes. This suggests that before there were cells, there probably were carriers of genetic information that eventually became viruses.

As the centuries rolled along and conglomerations of molecules began to develop the attributes of cells, complete with all the machinery needed to reproduce, instead of viruses serving the essential function of gene carriers, they began to "infect." Left behind in the wake of the evolutionary trend toward increasing complexity, viruses began to require the environment provided by cells in order to fulfill their own biological destiny. Infecting virtually every living thing, a virus carries with it, and often imposes on the cells it inhabits, a new genetic game plan.

When viruses occasionally jump from one species to another, their genetic material is reengineered to fit that of the new host, and the host cell may then pass the viruses' message to its offspring. The virus does this merely to survive; the fact that many viruses happen to damage living things—though of paramount importance to the infected cell, plant, or animal—is incidental. Virus may be merely an original mode of reproduction that's been superseded and replaced by more sophisticated assemblies of cells, but its ancient habits seem to be intact. It's as if viruses were left over from the primeval world and, like other ancient creatures such as sharks, spiders, and reptiles, fill our myths and bad dreams.

But the bite of an animal or insect, because it can be seen and felt, is less frightening than is an ill wind that blows in with an epidemic in tow. In many parts of the world where the germ theory is unknown, evil spirits are still blamed for illnesses caused by invisible entities.

Bacteria and Other Parasites

In a very real sense, the history of civilization is inextricably tied to the effects of invisible pathogens. When the population of the world was small, hunters and pastoral peoples lived in relative harmony with the multitude of microbes in their local environments. The advance of civilization changed this balance as people began redefining their place in nature.

Once permanent communities were established, a population density great enough to sustain and transmit diseases was achieved. The accumulated wastes of these large groups of people created pools of potential hosts for the parasites that formerly had been left behind by nomads. Domesticated animals also harbored organisms dangerous to people, and the needs of agriculture created environments in which disease-carrying insects, snails, and worms could breed.

Settled populations began to require goods and materials from outside their own areas, and trade between previously isolated groups of people brought novel pathogens into uninfected populations. New illnesses and deadly epidemics followed as smallpox, measles, syphilis, yellow fever, and a multitude of worms were scattered along trade routes.

Then disease began to be a major factor in social organization as it debilitated or decimated the people, rendering them unable to work and leaving them subject to incursions from outside. The need to control illness created central governments to carry out sanitation measures and quarantines and to deliver health care.

It was not until the middle of the nineteenth century that the health of the western world began to change for the better. Before vaccinations or antibiotics, before there was more than a rudimentary knowledge of what causes diseases or ways to treat most of them, people started to become healthier. In other parts of the world— now called the developing nations—such improvements have yet to be made and patterns of disease mirror those we once lived with.

The dramatic improvement in health rested on several significant social changes. Enough food was being produced to prevent undernutrition and the susceptibility to disease that accompanies it, sanitation systems disposed of human wastes, and clean water halted the spread of microbes from contaminated rivers and wells. Freed from constant disease, the inhabitants of the developed world could expect to survive childhood and have a long life in which to plan and carry out a continuum of projects unrelated to simple survival. As absurd as it sounds, if we couldn't flush toilets, western civilization as we know it wouldn't exist. Only in recent years has medical science played a significant role in health, largely in the form of vaccination to prevent illness.

Older readers may remember the years before World War II, when infections ran their course without more than token treatment. A contaminated wound could easily cause gangrene and require the amputation of a limb to save life. Children died from fulminating diarrheas and bacterial pneumonias and meningitis, tuberculosis, rheumatic fever, diphtheria, venereal diseases, and infestations of worms. All these diseases were virtually unmanageable, as most still are in the Third World.

While viruses are now our major concern insofar as transmissible, infective diseases go, in Africa, India, and other developing nations, other classes of pathogens still cause the premature death and debility they once did in the west. Because there is no clean water to drink or ways to render wastes harmless, because of limited access to vaccines and antibiotics, and because there is little understanding of the germ theory of disease, most death and illness on our planet are still caused by the bacteria and other "giant" parasites that were first revealed in a series of technological breakthroughs that began in the 1600s.

The First Look at the Invisible Domain

In the 1600s, a jack-of-all-trades and amateur maker of optical lenses, Antonie van Leeuwenhoek, fabricated a one-lens microscope that brought into sharp relief a wealth of previously unseen organisms. They were recognized by Leeuwenhoek as being forms of life, but their association with disease was beyond his training or interest. Common sense and observation had long suggested that diseases pass from one person to another, and fantastic theories such as noxious ethers and the spontaneous generation of life had been put forward to explain how this might happen.

But while the Dutchman's microscope made it possible to see bacteria, 200 years passed before the technology evolved to develop cultures in which to grow individual microbes and to show which one was responsible for a particular disease.

In the 1800s, first Louis Pasteur and then Robert Koch made a series of discoveries that set microbiology on its present course. The

innovative observations and techniques of these two men included
ways to grow colonies of cells and watch the effects of various microbes on them. Almost immediately, great advances in theory and
technique followed. Before the end of the century, the tuberculosis
bacillus and the germs that cause typhoid, diphtheria, tetanus, streptococcal infection, gonorrhea, pneumonia, meningitis, and staphylococcal infection were identified, although the agents that caused
them were not visualized. These invisible entities could pass through
filters that strained out the larger disease-causing germs, and their
presence was inferred through their ability to cause illnesses when
injected into animals or rubbed on plants.

Some of these entities were viruses and others were specialized
forms of bacteria, but it was the ordinary bacterial infections that
primarily concerned medical researchers. To understand what viruses are, it's important to know what they aren't.

What Viruses Are Not: Bacteria

The bacteria that infect and damage plants and animals constitute a minute fraction of the enormous number of microbes, many
of which play important roles in processes that are central to life.
Some bacteria convert nitrogen and carbon dioxide into usable organic forms; others are responsible for fermentation, making it
possible to produce beer, antibiotics, and the holes in Swiss cheese.
With an amazing ability to adapt to virtually every ecological niche,
bacteria live in the salty Dead Sea, flourish under the tremendous
heat and barometric pressures of the ocean depths, and thrive in
hot springs and subfreezing temperatures. Some bacteria live only
in the presence of oxygen, while to others this gas is a poison. The
shapes and sizes of different bacteria vary as widely as do their
habitats, with some as small as the largest virus—the poxvirus—and
others a hundred times larger.

Like all living entities, bacteria come with their own DNA and
independent metabolisms. While they require a hospitable environment in which to divide and grow, most do so without utilizing the
chemical machinery of the host. In other words, unlike viruses,
which must take over a cell's metabolism and reprogram it to produce viral offspring, a bacterium is fully equipped to carry out all

its life functions. Reproducing geometrically by division, most bacteria that cause illness do so only when their number reaches a critical mass. Others, such as the bacteria responsible for diphtheria, tetanus, cholera, botulism, and staphylococcal food poisoning, cause disease by releasing toxic substances, some of which are stunningly poisonous, outstripping even rattlesnake venom and strychnine in toxicity.

The most common bacterial disease is the one that causes cavities and periodontal disease. While all humans have more or less the same mixed bacterial population in their mouths, the sugary diet of affluent societies provides fuel for the microbes that act on tissues through their ability to ferment carbohydrates.

This same characteristic produces gas gangrene in wounds contaminated by *Clostridium perfringens*, in which fermentation produces lactic acid and gas. Compression of injured tissues by gaseous swelling prevents the oxygen carried by red blood cells from reaching cells in the wounded area, and the cells die. In severe cases of gangrene, several hours spent in a hyperbaric chamber of the kind used to prevent the bends experienced by divers increases the oxygen content of blood serum enough to interrupt bacterial multiplication. After this, the human immune system, with the aid of antibiotics, can finish the job.

Tetanus, or "lockjaw," which is routinely prevented by vaccination in early childhood, still kills hundreds of thousands of Third World newborns through umbilical cord wounds. The toxin released by *Clostridium tetani*, which live in soil, interferes with the transmission of nerve impulses, creating prolonged muscle spasms which may cause the jaw to "lock" closed or paralyze the muscles needed to breathe.

Most cases of pneumonia are caused by bacterial agents, with the overwhelming majority being due to streptococci. Legionnaires' disease, which made 182 people ill and killed 29 others in Philadelphia in 1976, was finally found to be caused by a new bacterium that spread through the air-conditioning system of the hotel in which the attendees of a meeting were staying. Subsequently, the same agent has been found in water storage tanks in several hospitals.

Bacteria continue to plague the population with urinary tract infections and with diarrhea caused by shigella, campylobacter, cholera, *E. coli*, and most often salmonella, which frequently contaminates poultry and egg products.

All but a few of the multitude of bacteria found in nature are independent organisms, able to live and multiply in an environment that suits their requirements. One small class of bacteria—the rickettsias—were long thought to be viruses because they invade the blood cells that line the heart and blood vessels. Carried and transmitted by fleas, mites, ticks, and lice, these bacilli cause classic typhus fever and other less virulent diseases, such as Rocky Mountain, spotted, and Q fevers. But rickettsias, like their bacterial cousins, are susceptible to antibiotics.

Despite modern advances in preventing and treating bacterial infections, many continue to have their way with humankind, and those that are sexually transmitted are the most common communicable diseases in the world. For instance, gonorrhea, which was described in 3500 B.C. in Egyptian papyrus scrolls, causes 2.5 million new infections a year just in the United States. Syphilis, which is caused by a spirochete (a slender, spiral microorganism), can be a fatal or disabling disease that damages the brain, heart, and muscles.

Syphilis was long considered as "shameful" as AIDS. It was only in 1936, when the surgeon general, Thomas Parran, wrote *A Shadow on the Land*, that syphilis came out of its closet and began to be managed in a public health setting. Almost exactly fifty years later, another surgeon general, Dr. C. Everett Koop, performed the same service by "legitimizing" AIDS, showing that scientific progress often fails to be matched by corresponding changes in public attitudes.

Other Infectious Agents

A class of pathogens called mycobacteria have characteristics that seem to fall halfway between those of bacteria and those of viruses. Far more resistant to chemical agents than are bacteria and able to hide in cells as viruses do, these pathogens cause tuberculosis and other lung diseases as well as leprosy (which, despite its dread connotations, is extremely difficult to transmit).

The advent of AIDS has focused attention on the mycobacteria as well as on various fungal infections, such as histoplasmosis, that seldom cause illness in industrialized nations. These esoteric parasites, which are usually held in check by a competent immune system, are responsible for many serious opportunistic infections in people whose immunity is lowered. The same is true of the infectious agents called protozoans, which cause amebiasis and giardiasis, two kinds of diarrhea that affect much of the world's population, primarily in areas where the quality of water is marginal or poor.

Malaria is the most important protozoal disease and is still a significant cause of death and debility in underdeveloped tropical nations. Until the 1950s it was a problem in the western and southern areas of the United States as well. An effective antimalarial vaccine would be an even more important public health measure than the smallpox vaccine has been, but the malaria parasite's ability to develop resistance to treatment—as well as the different forms it takes during its life cycle within the infected body—makes this an iffy prospect.

Antibiotics

Bacteria and many other microbes are independent life-forms with cell walls and membranes that are exposed to the internal environment of the body as they circulate in the bloodstream. There, the integrity of their protective outer walls and many of their metabolic processes can be disrupted by antibiotic drugs. The major action of antibiotics is to suppress the multiplication of disease-causing microbes. Because it is usually the sheer numbers of microbes rather than their mere presence that cause the symptoms of disease, when the number of bacteria or other microbes is limited, the individual's immune system is able to handle them.

"Although man can build a better mousetrap, nature always seems to build a better mouse," a textbook on infectious diseases reminds us.* When one remembers that the first effective antibiotic —penicillin—is simply a mold found in the environment, one realizes that microbes have been exposed to and have found ways to deal with natural antimicrobial substances throughout their evolutionary history. They respond in the same tricky way to those pro-

*G. Youmans, P. Paterson, and H. Sommers, *The Biological and Clinical Basis of Infectious Diseases*, 2nd ed. (Philadelphia: Saunders, 1980).

duced or purified in a laboratory. Almost as soon as penicillin was put to use in World War II, resistant strains of staphylococcus appeared.

Resistance to drugs develops when a few microbes are able to adapt to an antibiotic and thereafter multiply in the presence of the drug. Although drug manufacturers claim to be producing "new" antibiotics, these are merely modifications of existing drugs. In fact, no really novel class of antibiotics has been developed since the early 1960s.

Medical advances in and of themselves have brought a host of new situations into the health picture. For instance, organ transplants are effective only when the immune system is blocked to prevent the rejection of foreign tissues, but this allows ordinarily innocuous agents to proliferate. Because we live longer, chronic, genetic, and degenerative conditions have become a major focus of modern medicine. Many of these conditions are known or suspected to be caused by viruses. Although the role of viruses in most of these illnesses remains to be proved, it will eventually become apparent that the way viruses infect cells can bring about changes that have both short- and long-term effects on health.

2

The Process of Infection

What Viruses Are

Viruses are neatly described in scientific shorthand as unique obligate intracellular parasites. Unique they certainly are. *Obligate* means "without alternatives," with no choice other than to replicate as they do because they lack the equipment needed to reproduce on their own. *Intracellular* means "within the cell."

A virus is essentially a packet of genetic information surrounded by a protein covering. All viruses have a kind of covering called a capsid to protect their innermost core, and some have an additional envelope made up of proteins and fat. The molecules on this outer covering and those on the cell surface dictate the specific cells the virus can recognize and infect.

This specificity explains why relatively few viruses cause disease in different animal species or in both plants and animals. Although in the laboratory many viruses can be adapted to grow in the cells of experimental animals such as mice, creatures in the wild are seldom susceptible to diseases caused by particular pathogens. Only a relatively few viruses, such as those that cause rabies and influenza, are dangerous to totally different species.

Encounter with the Host

The encounter between a virus and its potential target starts when the virus reaches the appropriate cells of a susceptible creature. For instance, laboratory workers handling a virus that causes stomach disorders are in no danger if they breathe air containing that virus. They are, however, at risk if the virus is transferred from hands to mouth and thus to the gut. One can ingest viruses that have no chance of living in the stomach's acid environment but that can cause measles or flu if they are inhaled.

The hepatitis B virus targets liver cells; human immunodeficiency virus (HIV) goes for cells with the molecular marker called CD4; poliovirus infects stomach cells first and causes paralysis only if it reaches the central nervous system; and the hundreds of rhinoviruses that cause the common cold can be grown only in cultures of cells taken from the outer epithelial (skin) layers of the nose and throat. Unlike the herpes simplex and vaccinia viruses, both of which erupt on the skin, rhinovirus infections are not supported by skin cells.

Other very specific conditions also have to exist within the host. For example, rhinoviruses die when they are exposed to temperatures higher than those of the throat and nose. The heat in the lungs prevents cold viruses from replicating in those cells to cause pneumonia. Thus, if you have a cold that seems to be turning into a congested chest or flu, another virus or a bacterial agent may be involved.

Many viruses are particular not only about which cell type they infect but also about the state of the cell. If actively dividing cells are necessary, the activity provided by an ongoing infection with another microbe may be the key. Alternatively, the rapidly dividing cells of an embryo may fill the bill. Thus, the age at which an animal is exposed to a virus often makes the difference between a lethal versus an innocuous infection.

Some viruses, such as the rabies virus, are uniformly virulent, while others, such as the rhinoviruses, never are. But this difference sometimes depends on the virus isolate: Within some virus families, minute differences between isolates (strains) determine the amount of damage done to the host.

In short, it is the interaction between the surfaces of the virus and the potential host cell that determines whether infection will take place. While the outcomes of viral infections are predicated on many variables, a fairly routine cascade of biological events is initially set in motion.

Penetration and Creation

Having found cells displaying molecules that correspond to those on its own surface, the virus attaches to these by a process called adsorption. Depending on the type of virus, it may fuse with the cell's surface membrane and then release its contents into the cell. If the virus is naked, that is, lacks an outer envelope, it may actually burrow through the cell membrane.

One of the simplest viruses is the T4 phage that infects the *Escherichia coli* bacteria that populate the stomach. Topped by a twenty-sided capsid containing its DNA, the T4 phage has a cylindrical tail and long spindly "legs" like those of a spider. When the phage reaches the appropriate spot on the bacterium, the tail contracts, forcibly injecting its DNA into the *E. coli* through a tail tube driven through the bacterium's cell wall.

Remember that this takes place on a scale in which a short string of molecules can define both shape and function. Compared with the bacteria they infect, most phages are barely as large as a dot made by a sharp pencil would be in this paragraph. Even the largest of all viruses, the one that causes smallpox, looks like a postage stamp on a business-size envelope when superimposed on a bacterium.

Once the virus has penetrated the cell wall, its coat is removed by chemicals in the cellular cytoplasm. At this point a kind of eclipse takes place during which the virus can't be found with any laboratory technique. In one respect, the naked bits and pieces aren't a virus at all but elements which, when combined with the assembly-line capabilities of the accommodating host cell, begin constructing new virus.

Life depends on the ability to create and break down molecules of sugars, fats, and proteins. Sugars and fats are made up of carbon, hydrogen, and oxygen atoms; add nitrogen and you have protein. Proteins (the word means "first place") make up half the dry weight of a human body, and there are more than 100,000 different kinds.

Every cell contains thousands of proteins, the manufacture of each of which is directed by a gene. Viruses likewise have genes that direct protein synthesis, but without the metabolic functions or chemicals that would allow them to replicate on their own, they are obliged to use those supplied by the host cell.

Once infected, the cell becomes a factory behind whose walls the working parts—largely mysterious metabolic events—are hidden. Except in the case of the poxvirus, which uses only the cell's cytoplasm to replicate itself, the genetic material of the stripped-down virus enters the nucleus of the host cell and inserts its DNA into the DNA of the cell.

Now controlled by the virus, the cell's DNA is *transcribed* into messenger RNA, which contains the virus's genetic material. The RNA "message" then returns to the cytoplasm, where it is *translated* into proteins dictated by the virus. Viruses are able to do all this with a surprisingly small amount of genetic information. The relatively large poxvirus has enough genetic material to control the production of 200 to 300 proteins, while the tiny picornaviruses can direct the synthesis of only a few strings of amino acids.

The AIDS virus, HIV, so far is known to need a combination of at least fourteen proteins of its own plus the chemical talents of the infected cell to carry out its entire replication cycle. Two of the genes of HIV regulate "on" and "off" messages that tell the virus when to begin and end various processes. Another gene carries the program to make the reverse transcriptase molecule, an extra step that HIV, being a retrovirus, requires to translate its RNA to DNA. Yet another HIV gene directs the assembly of the new viruses' envelopes.

The success a virus has in productively infecting a particular cell depends on many factors. If, for instance, a cell is already infected with another type of virus or a defective viral particle, this may interfere with the newcomer's ability to assemble the chemicals it needs to reproduce.

Depending on its particular habits and requirements, the virus may remain within the cell or may kill the cell as it exits; alternatively, the newly formed viruses may be released slowly without causing cell damage. Most of this depends on the kind of infection a virus is programmed to produce. Some viruses are very straightforward, even dependable; others can play unexpected, dangerous tricks.

Types of Infection

Most readers will recently have had a "cold," which is just a collection of symptoms caused by any one of several hundred viruses. Many colds are caused by the rhinoviruses, of which there are 100 or more and which move from person to person by means of nasal secretions on the hands (not by sneezing, so prevention involves hand washing).

Colds may also be caused by coronaviruses, coxsackieviruses, or adenoviruses, but since half of all viruses that affect humans can cause the upper respiratory symptoms of a cold, a number of different categories of virus may be responsible for the few days of sneezing, stuffed runny nose, general malaise, and scratchy throat. A cold runs its short course without serious symptoms, so except for research purposes, it doesn't much matter which virus caused it.

Colds and other viruses that cause an immediate set of symptoms and then are eliminated are called *acute* infections. These viruses cause the minor kinds of illnesses most people think of when they visualize viral diseases. Acute infections can also be deadly, but whatever the outcome, the virus infects specific cells, and the symptoms of disease follow quickly. Many symptoms are caused by the death of infected cells and the body's response to this event. But natural selection favors any pathogen that doesn't kill its host, so *persistent*, as opposed to *acute*, infections are the ticket to a virus's survival.

Persistent infections can follow one of several courses. They may become *chronic*, as happens with the cytomegalovirus (a herpesvirus) when it infects infants. While a chronic virus may not kill infected cells, it can interfere with some of their functions. Chronic infections are thought to cause a number of diseases that involve constant but fruitless attacks by the host's immune system, as happens in many autoimmune disorders (Chapter 15).

Viruses that cause chronic infection can be cultured in the laboratory, while those that cause *latent* infections cannot be found. Latent viruses are able to evade the host's immune system, just as they evade the best efforts of scientists. Although retroviruses use this mode to remain persistent, it is the herpesvirus family that makes a specialty of latent infections. The herpesviruses have one thing in common: They become persistent and can be reactivated.

Herpesviruses

When genital herpes began to attract attention a decade ago, the American public became generally aware of persistent viral infections for the first time. Herpes simplex II is responsible for periodic eruptions of genital sores, while herpes simplex I causes mouth lesions.

Another herpesvirus is known by two names. When the virus causes chickenpox in the young, it is called varicella; when it is reactivated many years later to cause shingles in the elderly, it is referred to as zoster. Someone with a long-forgotten case of childhood chickenpox may come down at age fifty with the typical chills and fever of an infection. A few days later, clusters of painful vesicles, or pox, pop out around the trunk of the body. The virus, which had been sequestered in nerve cells, has become reactivated and erupted through the skin to cause what is called shingles. What is now in effect another case of chickenpox can be transmitted to a susceptible person.

Cytomegalovirus (CMV), with which most of the adult population is latently infected, can cause swelling of the lymph nodes and the malaise that is mononucleosis. But except for immunocompromised persons or infants, active CMV infection is relatively rare (Chapter 8). Epstein-Barr virus, another of the herpes family, is best known as the cause of mononucleosis but also has been implicated in various other diseases, including several cancers.

The last known member of the herpes family, human herpes 6, was discovered in 1986 at the National Cancer Institute by Zaki Salahuddin. This virus has been identified as the cause of rubella (German measles), may or may not be the cause of some cases of chronic fatigue syndrome (Chapter 18), and may be a cofactor in AIDS.

When one looks at these longtime fellow travelers of humankind, especially in the light of new technology, one begins to see that many of them also have different and devastating effects on infected people when they become persistent. Even infections caused by some of the viruses associated with common, acute illnesses can result in persistent disease states.

Unfamiliar Habits of Measles

The measles virus is an example of a familiar virus that plays many unpleasant tricks. It is a member of the family of RNA viruses called paramyxovirus—which includes the mumps virus—and is related to the virus that causes distemper in dogs. The measles virus is incredibly easy to transmit. A child can catch measles by breathing the air in a doctor's waiting room two hours after an infected child has left. The infection results in high fever, sore throat and cough, and the appearance of the familiar rash. In otherwise healthy children, the disease resolves within a week or so and lifelong immunity is established. That's the ordinary sequence of events, but because the virus causes a transient suppression of the immune system, some patients are at risk for tuberculosis, pneumonia, and other bacterial infections to which they are exposed or with which they are already infected.

An acute measles infection makes a huge demand on the body's vitamin A, abundant supplies of which are stored in the liver and released as needed to maintain the health of the skin, hair, and eyes. However, many Third World children are deficient in vitamin A and can become blind and sometimes die from the total depletion of this essential vitamin during measles infection. This is a dramatic example of how host factors influence the outcome of an infection.

Unexplained, however, is why 1 person in about 800 infected with measles experiences encephalitis, an inflammation of the brain that causes convulsions and coma. Some recover within a week or so, but others may be left with serious impairment of the central nervous system, and some die. Whether encephalitis is caused by a mutant strain of the virus or by the immune system's attack on it isn't known.

Another rare outcome of measles is subacute sclerosing pan-encephalitis (SSPE), a deadly brain disease which occurs months to many years after acute measles infection in infancy. SSPE signals its presence with deteriorating mental abilities, convulsions, and abnormalities in motor function. In this instance, a virus that ordinarily causes an acute infection has become latent and behaves like a *slow* virus, which takes many years to manifest its presence.

Dr. Michael Oldstone, the foremost researcher into molecular mimicry (in which the receptors of a virus or of the antibodies against it may bear a close similarity to the host cells), theorizes that in SSPE, antibodies that mimic host proteins may bind to the infected cells so as to hide the signs of infection from the immune system. With nothing to recognize as foreign, the immune system simply ignores the infected cells, and the field is left open to the virus. (Antibodies are protein molecules produced by white blood cells when these cells encounter a pathogen.)

In effect, then, the measles virus may cause the familiar acute infection or an immune system response that is dangerous to the patient, or it may become latent and cause death years later. The fact that we are latently infected by numerous ordinary viruses as well as other germs has been highlighted by AIDS. The opportunistic infections that kill in AIDS are those with which the HIV-infected person has previously lived in harmony.

Escape Artists

Some viruses use their ability to mutate rapidly to evade immune system attacks. A fascinating example of this evolutionary game was played out in Australia in the 1950s. When European rabbits were foolishly brought to that continent, they bred at an alarming rate. Without natural predators to control their reproduction, the animals became a serious agricultural pest, and in 1950 an American virus lethal to rabbits—the myxomatosis virus, a distant relative of human smallpox—was introduced to reduce their numbers. Ninety-nine percent of the rabbits were killed in the first wave of the infection.

However, the few rabbits that recovered were thereafter immune, and in areas where the resistant rabbits bred, the mortality rates of their descendants fell dramatically. Within seven years, only 25 percent of exposed rabbits died. In some parts of the country, attenuated (changed or weakened) strains of the myxomatosis virus had come into being, allowing the virus to survive in a population of rabbits the majority of which became resistant to its effects.

The changes wrought by the myxomatosis virus on the rabbits, and those wrought by the resistant rabbits on the virus itself, are an example of the survival of the fittest. In this case, both virus and host were sufficiently fit to accommodate and survive.

This parallels the experience of the Native American populations that came into contact with unknown European viruses such as smallpox, which killed enormous numbers of their people. Their survival as a race depended on those who recovered being able to develop immune system defenses against the various pathogens and to pass those defenses on to their children. This is virtually the same mechanism that encourages the development of resistance to antibiotics by many bacteria.

However, there are other, far more exotic agents that cause infections, and these agents were first found not in humans but in plants. Although as far as is known there is no cross-species infection between human and plant, the similarity of some of the agents that infect both organisms argues for an ancient common ancestor. The small incomplete agent that causes delta hepatitis was preceded by the discovery by plant physiologist Theodore Diener of an entity that broke all the known rules of the game.

Dubbed viroids to distinguish them from viruses, these agents lack any kind of covering to protect their genetic material. Without even genes to enable them to use the machinery of infected cells to reproduce, viroids are nonetheless able to cause infection. These infinitesimally small strings of amino acids shed light on the ancient past of the planet; at the same time, they hint at a future in which the differences between animal and plant cells will be reconciled to serve the needs of both life-forms.

3

The Rolling Circle:
Discovering the Viroid

A sculptor who carved marvelous elephants from wood, when asked how he could make them so lifelike, answered in amazement, "Why, I just cut off everything that isn't elephant!" That's more or less the way Dr. Theodore Diener discovered that there is such a thing as a viroid. The discovery of this previously unsuspected type of infective agent, although it was found in a plant, led to a new way of thinking about human viruses.

Even without a link to human disease, Diener's proposition of the existence of a viroid was disturbing to scientists. It contradicted long-held beliefs and forced a rethinking of the whole field of virology. Diener proved that the virus, with its strand of genetic material enveloped in a protein coat, which was thought to be the smallest entity capable of transmitting disease, is in fact a large and complex organism compared with a viroid.

The primitive way in which a viroid replicates points back in time to what viruses might have been like before there were animal cells to infect. The existence of the viroid has also made sense of similar minute agents that can combine with other viruses to cause very severe diseases, such as delta hepatitis. But Dr. Diener could have guessed none of this when he began looking at plants to see why some were healthy and others were not.

"This plant is from Costa Rica, and it has a viroid in it that we accidentally discovered," Dr. Diener told me, casting a wary eye at the ordinary-looking houseplant dangling from the basket in his office window.

"The viroid doesn't cause any symptoms in this plant, but if you put it into potatoes, you get a very serious disease," said Diener. "I don't know why we haven't had that epidemic yet: I can easily see how some farmer's wife takes home a plant like this and puts it on the porch. Then the cat brushes against it, jumps down, and goes into the potato field outside and. . . ." Diener momentarily muses on a worst-case scenario, the vision of acres of stunted potatoes, then continues:

"The plant pathologist scratches his head trying to figure out where this comes from but would never suspect the ornamental plant." Diener would: He's spent most of his sixty-eight years figuring out what causes plants to become diseased.

When he embarked on his career as a plant pathologist, if what was causing problems with crops wasn't a bacterium or a fungus, it was presumed to be a virus. Though no one had actually seen a virus, their existence was assumed. Virus was the pathogen too small to see, an entity so minute that it would pass through the porcelain filters that screen out bacteria and fungi. It clearly was able to transmit disease.

This had been shown at the close of the nineteenth century by botanist Martinus Beijerinck. Passing juices squeezed from diseased tobacco plants through porcelain filters to remove all cells and known infective agents, Beijerinck found that the clear fluid still transmitted the disease to healthy tobacco plants.

Beijerinck conducted various experiments that seemed to show that the fluid itself was infectious, and he published a report describing what he called a virus. What he meant by the term wasn't the submicroscopic entity we know today but rather described what to the botanist was a "primitive form" like "a flame borne by the living substance." Beijerinck had no way to see the rod-shaped tobacco mosaic virus that had easily slipped through his filters. As physical entities, viruses were to remain invisible until the development of the electron microscope in 1939.

The first virus to be photographed through the electron microscope, which uses a beam of electrons rather than light for illumination, was the same tobacco mosaic virus. Though the picture of the slender rod was infinitely clearer than were the specks seen under the light microscope, knowledge of the internal assembly of the pathogen was many years down the road.

When Diener migrated to the United States from Switzerland in 1949, virology was still very much a case of working from inferred evidence. Filterable agents were known by then to be viruses, but their effects rather than their actual presence or structures were all experimenters had to work with.

Diener's first job in America was to investigate the viral infections of fruit trees in an outlying agricultural station in Washington with "not much of a lab." Not wanting "to go another round in the field with a budding knife," Diener began looking for a challenge. He decided to initiate a project to discover the differences between healthy and infected trees.

"I found something very interesting, an unusual amino acid that accumulated in diseased trees," he explained. This immediately established him as the world's foremost expert in the physiology of virus-infected plants, as he says, because nobody else knew much about it. In any event, he "didn't want to work on something that was already known."

In 1957, America's interest in science, so great during World War II, had fallen into the doldrums. Then the Russians launched *Sputnik 1*. Faced with this challenge, the country realized how far behind it had slipped in science, and a massive amount of money was immediately put into basic research projects, including, as a kind of afterthought, the creation of twenty new laboratories for the U.S. Department of Agriculture.

"They wanted someone to work on symptoms, and of course, who else was there?" Diener asks rhetorically.

He and the group assembled at the Agricultural Research Center in Beltsville, Maryland, where he remains still, found a lot of new things, but "eventually I stumbled into this viroid problem," Diener explains. "I didn't know what it was then. For quite a long time it was a sort of Saturday afternoon experiment that I thought probably wouldn't lead anywhere or would have a trivial explanation. But it became more and more interesting. Eventually that became my only area of work."

What Diener had found was a completely novel form of infective agent—a viroid. It was a naked strand of genetic material, lacking even the protein coat of most viruses, but it was able to twist potatoes into gnarled, unsalable gnomes or to spoil a field of tomatoes.

While it is hard for nonscientists to understand the importance of such a discovery, knowledge of the viroid's existence raised the possibility that similar types of agents may play important roles in human disease. In addition, the viroid's mode of replication expanded the intellectual framework in a way that allowed other researchers to look for agents whose existence, before Diener, wasn't even suspected.

Seven years later, in 1964, after he had "plugged all the loopholes," Diener finally published a report of his work on viroids in a scientific journal.

"There was incredible skepticism! For several years I was invited everywhere by the high priests of molecular biology, who said, 'Mind you, Diener, I haven't read one of your papers but I know such things can't exist.'" Diener recalls that the molecular biologists in particular were very uneasy with evidence obtained by rubbing something on tomato plants. "But we have to be skeptical in science. We have a lot of crackpots, and if you run after all the crazy people who come up with things, you couldn't go anywhere."

The high priests of medical research had apparently forgotten that the great discoveries about and around viruses had been made by chemists, botanists, physicists, and crystallographers dealing with viruses that infect plants.

Genuinely novel ideas in science are always initially met with disbelief by people working in the same area. The announcements that finally appear in the news are like the meal you order in a restaurant. The meat has been seasoned, neatly patted into shape, and then cooked and served with a potato and vegetables. But if you thought about the activities that made this simple fare possible, you'd have to go back step by step through many elaborate processes. Years of grazing were needed for the cow to accumulate fat and muscle, culminating in a trip to the stockyard and the violence of the slaughter. The potatoes had to be planted, grown, dug, washed, peeled, and cooked and the vegetables had to be fertilized and sliced before the final product could arrive at the table decorated with a bit of parsley. The analogy between this preparation and the scientific activities that result in "breakthrough" announcements often include the slaughter.

To come to a new conclusion and to be able to prove it means that the older theory it necessarily displaces was wrong or incomplete. The kind of significant discovery Diener made results from a long process that is fraught with danger, not the least of which is that it may be wrong. The scientific community, all the members of which are competing for scarce funding, lies in wait to pounce on mistaken assumptions and, if it can, to refute results. As nearly as possible, science must be built on layers of testable facts. A scientist is free to hold any theory he or she believes, but Harvard's Steven J. Gould makes clear the distinction between theory and fact: "Facts are the world's data. Theories are structures of ideas that explain and interpret facts." Diener had tossed into the hopper a fact that challenged existing theories.

The timing of a new discovery has to be right. The important fact that viruses can mutate had also been discovered in Beltsville in the 1930s, but "the man who did it—McKinney—was ridiculed for twenty years. At that time nobody knew what viruses were," Diener recalls. Thus Dr. McKinney and his important idea went nowhere in his lifetime. By the time Diener found the viroid, however, the establishment was ready for such an anomaly.

But it wasn't many years after establishing that there is such a thing as a viroid that Diener veered into error by suggesting that viroidlike agents were likely to be the cause of a neurological disease of animals called scrapie. "I published the paper in *Nature* saying this, but it turned out not to be true," he says. It was a plausible theory but a wrong one.

Manipulating the Unseen

In 1962, several researchers at the U.S. Department of Agriculture (USDA) lab found that the unidentified agent of potato spindle disease on which Diener was working could be transmitted to tomatoes. When they were ground up and placed in a solution, the tomato leaves in turn were also very infectious, as Beijerinck had shown almost 100 years earlier. Since that time, an important tool for extracting virus from fluid had come into being, the high-speed centrifuge.

Ordinarily you can spin a viral-infected solution at about 100,000 times gravity in a centrifuge, and the virus in it will form a pellet because it is heavier than the fluid. But no virus spun out of the ground-up potato-spindle-infected leaves even though the liquid itself remained infectious. At that point Diener entered the picture and began cutting away everything that wasn't elephant.

"I could show that it [the liquid] was very contagious in the field," recalls Diener, "and that it was spread by the knives used to bud plants."

The scientist had even dipped the contaminated budding knives in alcohol and then ignited the alcohol to kill any infective agent, but the knives could still transmit the disease. "There was something a bit unusual about this," recalls Diener, because all viruses then known were killed or rendered inactive by such drastic treatment.

Because he had to know which genetic material might be involved, an enzyme that breaks down DNA was added to the tomato-leaf-derived liquid. It still caused infection, so it wasn't a DNA virus. Other tests showed that there wasn't any protein or fat, substances known to constitute the envelope of viruses, in the liquid. The conclusion had to be that the virus lacked a coat or envelope (even the ordinary viruses termed naked have capsids around their genetic material). Because of its refusal to centrifuge, whatever it was had to be incredibly small.

However, there was a big intellectual problem with that idea. As far as anyone knew, an infective agent *had* to have a certain minimum amount of genetic material to be able to cause infection.

"It seemed plausible," Diener says, "to regard the infectious RNA as a coatless defective virus that multiplied in plant cells with the aid of a helper virus. In plants we have satellite viruses with a very small RNA, but they can't replicate by themselves and have to have what we call a helper virus." (The corollary in animal viruses is discussed in Chapter 13.)

A long search for another pathogen that could complete the hypothetical new, and presumably incomplete, agent was fruitless, as were efforts to show that the RNA was part of a larger RNA population that somehow could assemble itself from bits and pieces to form a single effective agent.

"At that time I couldn't see the agent as a physical entity. I could only show it by running it through all kinds of systems and taking individual fractions and putting them on tomatoes, which would then get the disease. So I determined all the essential properties of the agent without ever seeing it," Diener explains.

Until Diener's discovery of the viroid, scientists had been certain that they understood just how viruses infect cells. First, the virus's protein coat attaches to the host cell and dumps its own genetic material into the cell. Although dependent on the cell's ability to translate genetic information into action, viruses nevertheless bring with them instructions on how to do so. In other words, with the cell's chemical equipment to work with, the virus's genetic code directs the cell to produce one or more specific proteins required to replicate itself.

It is generally accepted that a virus's nucleic acid has to have a molecular weight (which is obtained by counting the number of atoms) of at least 1 million to take over a cell's genetic machinery and code for the production of viral proteins. Poxviruses, with a molecular weight of 160 times 10^6, can code for between 200 and 300 proteins, and picornaviruses (2 times 10^6) can code for perhaps 5 to 8. Diener's viroid—merely a strand of RNA—would turn out to have a molecular weight of only 130,000. But although they are too small to have the number of genes needed to instruct the cells they infect, viroids are able to replicate quite efficiently.

Visually, alongside the *E. coli* bacterium that inhabits the gut, the large poxvirus looks like a tugboat docked next to an ocean liner, while the tiny picornavirus is about the size of a porthole. A viroid is as small as a coil of rope on the tug's deck.

The life and health of a cell depend on its genes directing the activities that need to be undertaken. If you view DNA or RNA as enormously long strings of beads (amino acids) made up of sections of different colors (subunits called nucleotides) and animate this string to act like a gene, each color sequence will be at work on a different task.

For instance, the red sequence at one end of the string may carry the signal to start making protein, while the red one at the other end may be responsible for halting production. Other sequences are busy turning on and off various activities, such as controlling cell growth or producing other proteins. In between, blue lengths —the introns—may be editing themselves out of the DNA sequence, making it possible to splice a stripped-down version of DNA into RNA.

If a virus with its own genetic agenda is introduced into this orderly plan of organization, it may change the normal signals. The infection may be benign, may cause cell death and acute illness, or may create a latent infection that will surface at a later date. But when a virus infects a cell, the process becomes a bit like sex that results in conception: the bringing together in one individual—be it an animal egg or a virus-infected cell—of two separate sets of genes. That's the way scientists used to think about viral reproduction, but the viroid turns out to have its own unique reproductive device.

In 1971, after seven years of viroid research, eliminating what it couldn't be and determining what it had to be, Diener was finally confident that he had found a unique infective agent, a free strand of RNA about one-tenth the length of any known virus. After the publication of this description of the viroid, over the next few years the U.S. Food and Drug Administration lab, as well as researchers in other parts of the world, worked to determine the arrangement and number of the nucleic acids that make up the particle's genetic material. At the Max Planck Institute in Munich, Hans Gross described the newcomer as having only 359 molecules, which were arranged in a closed circle, a unique structure in nature. But how did it replicate?

"That's what's so exciting, you see. We're still working on that very hard," Diener says. "It has to have some sort of a signal which tells the plants, 'Come on, you replicate me!' because it doesn't do it itself. It's the host enzymes that are replicating, but why? We still have no idea."

Diener has been able to purify these host enzymes to make a growth system in test tubes. When the viroid is put into the growth medium, it serves as an active template which can make complete copies of itself, literally rolling out one viroid after another.

"When you bring in the circular RNA, an enzyme starts at a given point to make the opposite strand. It goes around, makes the whole thing, and then it comes back to the origin, and instead of falling off, it [the enzyme] peels off what has already been synthesized and goes around again. We call it a rolling circle," explains Diener. Because viroids, unlike viruses, introduce very little genetic information into host cells, it appears that the cellular enzymes alone are responsible for viroid replication, working in response to a signal introduced by the viroid.

Since the viroid was discovered, comparable agents have been identified as the cause of other plant diseases, such as cucumber pale-fruit, the stunt disease of hops, and cadang-cadang, which has killed millions of coconut trees in the Philippines.

As economically important as it is to understand the diseases that destroy our foods, it is those that damage ourselves that are of major concern. Recently, a virus particle that seems to combine the peculiar attributes of atypical plant and animal infective agents has been found to complicate hepatitis B. Called delta, it replicates in exactly the same fashion as do viroids, by rolling itself in a circle and peeling off copy after copy (Chapter 13).

While there may not be any give-and-take between human viruses and those of plants, at least not in the way we exchange illnesses, intriguing similarities between them are appearing. If evolution elaborates on themes the way it appears to, if it adds new capabilities to its growth systems, then the naked RNA of a viroid seems to be a likely jumping-off point for determining how entities with genes to share began their long trek up the evolutionary ladder.

"Viroids," Diener speculates, "may be living fossils, RNA molecules that have survived [and evolved] since their origin during the very early prebiotic stages of evolution."

Dr. Gail Dinter-Gottlieb has written in the *New England Journal of Medicine* that viroids may be "escaped introns," the edited-out parts of plants' genes. Moreover, she points to the similarities between the introns of human cellular RNA genes and viroids, which make it almost certain that the two have an evolutionary relationship. Dinter-Gottlieb suggests that "the question still remains as to whether viroids evolved from introns or whether both evolved from a common ancestor molecule."

Assuming a "mother" molecule that existed in plants before significant animal life appeared on Earth, when RNA rather than DNA was the primary genetic material, viroids suggest the way in which different genetic information was once passed between species. Moreover, viroids retain certain habits that are reminiscent of the way retroviruses such as HIV replicate.

In any event, the viroid and the plants it infects are marvels of simplicity compared with the interactions between a living body and the constant deluge of elaborate viral agents that seek to use it as their breeding ground. Perhaps because plants, unlike animals, have no immune system defense to complicate their parasites' life cycles, viroids are able to replicate with so few molecules. But to infect a human, a virus needs many molecular tricks to escape the elaborate immune systems that animals have developed to protect themselves from an environment swarming with submicroscopic germs.

PART TWO

THE HOST

4

The Internal Ecosystem

Two Hundred Years of Innovation

The gatekeeper of the Pasteur Institute, a middle-aged man named Joseph Meister, committed suicide rather than hand over the keys of Louis Pasteur's crypt to the German forces that occupied Paris in 1940. This futile gesture of homage to the great scientist acknowledged Meister's debt: The life he took had been given him fifty-five years earlier, when as a child he was the first recipient of Pasteur's vaccination against the deadly rabies virus.

Pasteur's vaccine of 1885 was the first to be developed since 1798, when Edward Jenner devised one against smallpox. Although eleventh-century Chinese physicians had used powdered smallpox scabs to protect patients against the disease, it was only in the early 1700s that this practice was brought to the attention of European scientists by Lady Mary Wortley Montagu, the wife of the British ambassador to Turkey, herself scarred by the disease.

While the fact that vaccines protect their recipients against smallpox and rabies was already known, the reasons they do so remained a mystery until the middle of the twentieth century. The same lack of understanding was in part responsible for the ease with which small bands of European explorers conquered the indigenous people of the Americas. There were other aspects of the white men that seemed magical to the Indians, not the least of which was that Indian myths had foretold that tall, pale people would someday appear on their shores. To add to the aura of power, the Spaniards appeared riding the first horses seen on the continent since pre-historic times. But the invaders carried a far more formidable magic, one that would have been as mysterious to them as it was to the Amerindians: Their blood carried viruses never before known on the continent.

As hard as it is to imagine, before contact with whites, the in-habitants of the Americas had been spared any epidemics and thus had no rational way to deal with the devastation that followed in the wake of the invaders. Within weeks, up to 90 percent of some populations became ill and died.

More fear-provoking even than the deaths must have been the Spaniards' seeming imperviousness. This led Indian and white alike to assume that the hand of God was implacably set against the native populations. Not surprisingly, millions of Indians were converted to Catholicism as the cultures of parts of the Caribbean, Mexico, Peru, and Guatemala crumbled under the mysteries of smallpox.

Between 1518 and 1525, when Pizarro invaded Peru, an esti-mated third of the original populations in those countries had died. It didn't stop there. The survivors of the pox were next inundated by waves of influenza, measles, and mumps, infections that also had been unknown in the Americas. Although the same diseases reg-ularly swept across Europe and killed thousands of people, the survivors were in effect immunized against subsequent infections. Europeans landed in the New World carrying pathogens to which they were immune; the Indians unfortunately were not.

However, it seems probable that syphilis, or the "Spanish pox," was brought from the Americas to Europe during the earliest days of exploration. Several lines of investigation make this likely. An-thropologists at the University of Massachusetts have identified signs of syphilis on the bones of hundreds of skeletons from the New World; similar markings have been identified in European remains only rarely and tentatively.

Also, syphilis seems to have appeared as an important disease in Europe only in the 1500s, when lethal outbreaks of the venereal infection were first reported. People contracting syphilis were ravaged, as often happens when a novel pathogen is introduced into an unprotected population. The corollary to this is the virulence of measles and mumps in the Americas, whereas in Europe these were illnesses of childhood.

Surprisingly, this transoceanic intermingling of diseases seems to have included malaria and yellow fever, which are generally accepted as having been unknown in the Americas before the middle of the 1600s. These two mosquito-borne diseases were the reason West Africa was called "the white man's grave" when, not many years later, Europeans began their colonial incursions. Indirect evidence that these diseases were imported into the tropical girdle of the New World comes from looking at the differences in the genetic traits of African and Amerindian populations. Africans, who had long shared their continent with malaria-infected parasites, developed the sickle cell trait as a biological defense against the red blood cell parasite, a trait that Amerindians don't share.

The first known outbreak of yellow fever surfaced in Havana and the Yucatán in 1648, killing the previously unexposed human and monkey populations. The Massachusetts Bay Indians were also primed for surrender to Europeans by the infections that moved south from a new French outpost in Nova Scotia. In 1616, a "great pestilence" swept through the area, and when the Pilgrims arrived three years later, they assumed the mantle of divine protection that had been ascribed to the Spanish.

The reasons for the apparent impunity of the Europeans to these killer infections would take 400 years from the time of the conquistadores to become clear. What had seemed to the Indians to be the mystical strength of the white man in fact was the ability of white blood cells to remember and combat infectious agents to which they'd previously been exposed.

The development of this natural immunity is imitated by vaccines which expose immune system cells to a pathogen that has been killed or attenuated. This principle was accidentally discovered by Pasteur when a stale culture of chicken cholera bacillus, instead of causing disease, protected the fowl against subsequent injections with fresh cholera cultures. Although the amount or state of the virus in a vaccine is inadequate to cause illness, it programs specific antibody molecules to rush to combat the agent.

Host and Parasite

That people who recover from some diseases thereafter are resistant to it is a commonsense observation that must have been made thousands of times over history. This knowledge was translated into a crude kind of smallpox vaccination in which material made of ground-up pox scabs was scratched into a small vein. Lady Montagu's advocacy of this practice began soon after her husband moved the family to Turkey, where she attended a "variola (as smallpox was called) party."

In 1718, the Montagus' son was vaccinated, and three years later, the daughter. The practice was widespread in the American colonies and England by the time Edward Jenner was born in 1749. As a country doctor, Jenner had the opportunity to observe that as a group milkmaids seemed to be protected from the disfiguring illness after contracting cowpox, a mild form of pox that infects cows.

Using the technique of the lay practitioners, in 1796 Jenner scratched powder made of ground-up cowpox scabs into the arm of a child and six weeks later exposed the child to material from a smallpox patient. There was no infection. After resisting the idea for several years, the medical profession finally accepted the procedure as scientifically valid and adopted it.

The reluctance with which his colleagues accepted Jenner's novel preventive is hardly surprising given the state of biological knowledge at that time. Jenner and the folk practitioners were far ahead of their time in that smallpox is an unusual disease that can be prevented by recovery from another, related disease. (It's interesting that the first disease to be prevented by vaccination is also the only one we have been able to virtually eradicate nearly 200 years later.)

In Jenner's day, infectious agents were just beginning to be seen under the low-powered microscopes that were becoming available. Knowledge of the human immune system was almost nonexistent.

Since the late 1800s steady advances in technology combined with discoveries about the properties of blood cells, immunoglobulins, interferons, and various cell control factors have been woven into a special study of the immune system. It is almost as if the research community had been readying itself to deal with the appearance of the granddaddy of immune system diseases, AIDS. Had

AIDS come even a few years earlier than it did, we would still be mostly in the dark about why young people are dying from the most ordinary microbes, agents that ordinarily have no significant effects on the health of the host.

The great killer of AIDS patients is *Pneumocystis carinii* pneumonia, which is caused by a type of fungus carried harmlessly by the majority of adults. Those with AIDS also suffer from common bacterial infections such as staphylococcus and streptococcus; from mycobacterium (the tuberculosis agent) and *Pseudomonas* bacilli; from the protozoan disease toxoplasmosis avian-intracellulare, which infects many animals and birds; and from various rickettsia (which have habits halfway between those of viruses and those of bacteria and are carried by fleas and ticks) and spirochetes as well as a panoply of viral infections. All these agents are handled with aplomb by a competent immune system.

By the time AIDS surfaced, physicians had enough basic knowledge to suspect that it was the immune-suppressive effects of the virus that were allowing the indigenous microbes to become activated. Comparable though less dramatic opportunistic infections had been observed in people whose immune systems were depressed.

Much of this experience had been gained since the beginning of organ transplantation in 1967, when South African surgeon Christiaan Barnard performed the first heart transplant. Because the heart or kidney of another person is a foreign tissue, to prevent the patient's immune system from attacking it, suppressive drugs have to be given, often for the rest of the recipient's life. Otherwise, a dangerous graft-versus-host reaction causes the organ to be rejected and the patient to die. Many of the opportunistic infections, as well as the cancers experienced by those with AIDS, thus become serious problems for transplant patients.

The immune-suppressive drugs—such as cortisone, prednisone, and cyclosporin—were being used to treat persons with autoimmune conditions such as lupus and rheumatoid arthritis. Such drugs dampen the body's immune system attack against its own tissues. These patients, too, became highly susceptible to opportunistic infections.

An additional problem that may arise with the use of immune-suppressive drugs is that many of the body's indigenous microbes are important in preventing infection by competing pathogens. For instance, when the normal microbes, particularly those in the gut, are killed by antibiotics, the way may be opened for colonization by dangerous bacteria or fungi. An attack of diarrhea that occurs while a patient is taking antibiotics often signals that this has happened. (Chronic changes in the bacterial flora of the gut are thought to be an important intermediary step in the development of cancer of the large bowel.)

In most instances, the immune system of a healthy person is able to limit the proliferation of an infecting agent either through a vigorous immune response or because specific defenses against the agent have been generated by prior exposure. Many of these prior exposures are too mild to cause symptoms, but cells programmed to remember them lie in wait in the blood.

Most "extracellular" parasites—this includes virtually all but the fungi and viruses—are ready targets for antibiotic drugs and immune system attacks. Moving about in the bloodstream, these parasites can be destroyed by scavenger cells, antibodies, and the disruptive actions of antibiotics on their outer walls. Unlike the viruses, which become chronic or latent by hiding their presence from cells of the immune system, bacteria and other parasites are usually eliminated completely. Only when the numbers of parasites are too great are antibiotics needed to temporarily hold the pathogens at bay while the immune system swings into action. But viruses are impervious to antibiotics, and it is largely the interaction between virus and immune system that dictates the outcome of an infection.

The Immune System: Ten Million Possibilities

"With our eyes we look out on the savannah and see the wildebeest herds; with our immune system we can sense the presence of wildebeest macromolecules in our gut or in our blood," writes Case Western Reserve University's Dr. Robert Kaplan in the *New England Journal of Medicine*. How does an immune system born and bred in Brooklyn or Berkeley recognize an animal from the East African plains? The molecules that identify the animal are strung in a way different from the way in which those of a person are strung, in an exclusively wildebeest sequence that is registered as foreign.

The immune system has the inherent ability to recognize 10 million configurations of molecules. Viruses, parasites, fungi, bacteria and their toxins; pollens, strange blood cells, and a host of human-made molecules are scanned, evaluated, and accepted or rejected. At the same time it's guarding against these external dangers, the immune system monitors cells of the self and mops up behind the minor accidents that occur constantly within our bodies. It also rides herd on cells in the process of becoming cancerous.

We're used to hearing about the immune system in terms of "attackers" and "defenders." Lymphocytes are described as soldiers confronting an enemy. While there is an element of truth in such simple images, they fail to take into account the behind-the-scenes complexity of immune system cells and their chemical communication networks.

Disease and dysfunction aren't merely the result of one type of cell losing to another. They are caused by an imbalance, a mistiming, a lack of harmony, a faulty reading of signals at the molecular level.

Ben Franklin's doggerel that begins "For the loss of a nail, the horseshoe was lost" traces the series of events that lead to the loss of a kingdom. Similarly, the cascade of immune system activities that allow, and sometimes even generate, diseases is as complex as determining what happened between the time the horse dropped its shoe and the kingdom's eventual downfall. While the impact of a single individual act may be small, its consequences can be central to the final outcome.

An adequate immune response depends on many factors, including an individual's current health status, which may decline in the presence of an infection or from malnutrition or perhaps because of that individual's particular genetic makeup. Sometimes the information gets scrambled when an aging body's machinery accumulates small but ultimately critical errors.

Modern industrial and agricultural chemical pollution and increased solar radiation created by ozone breakdown in the atmosphere are also likely to play a part in speeding or slowing cellular activities. These insults may even change the normal programming of cells to make them more receptive to mutations and damage, encouraging cancers, autoimmune diseases, or allergies. But external influences on individual health aren't new. They exist in the environment just as natural antibiotics do.

Some people are born with an extraordinarily well balanced

chemistry which allows them to tiptoe through 100 or so years of life without succumbing to acute infections or developing chronic diseases and cancers. Many other people are shortchanged by their genes and are programmed from conception for health problems such as diabetes and hemophilia or atherosclerosis (the accumulation of fatty material inside major blood vessels) or more esoteric genetic mutations that make a person unable to digest wheat products or create the risk for cancer.

For millions of people, the immune system is a double-edged sword, protecting them but sometimes causing disorders that range from trivial to terrible. Allergies are misguided immune assaults on harmless substances such as nuts or bee stings; in rare cases, an allergic reaction is so overblown that it freezes the autonomic nervous system, which controls essential functions such as breathing, and thus causes death. Similarly, many widespread crippling diseases of people and animals are created by misidentification on the part of the immune system of self for other, of friend for foe.

Every illness or infection, as well as good health and resistance to disease, is controlled to some degree by the unique immune makeup of the individual. To understand this, we have to look at the elaborate communication network of chemical messages. This network is based both on a general scheme and on the individual differences conferred by heredity. There are about 60 trillion cells in the human body, almost every one of which carries the genetic code—the genotype—of the individual.

Elements of the Immune System

A living being is an assembly of organic molecules that have come to function together over evolutionary time in a relatively cooperative venture. The immune system itself is made up of organs located throughout the body. Because they are involved with the development and deployment of lymphocytes these are generally called lymphoid organs. They include the bone marrow, spleen, thymus gland, and lymph nodes as well as the tonsils, appendix, and clumps of cells in the small intestines called Peyer's patches.

The lymph system is linked in much the same way as the blood system is, with a network of vessels connecting the lymph nodes, collections of which are easily felt in the groin and armpits. Threaded through the body like rivulets, the vessels of the lymphatic system merge into increasingly larger tributaries through which lymph fluid flows, finally to empty into the bloodstream, where it is disposed of. The clear lymph fluid carries many cells and particles; as it drains through the lymph nodes, it picks up and deposits these and other substances.

The molecules that constitute the proteins from which life is created group together to form the cells of many different body systems: blood, cardiovascular, nervous, gastrointestinal, connective tissue, and so on. The great differences in the functions of cells dedicated to various organs depend in part on which of their repertoire of activities happens to be needed. In the fetus, for instance, cells must be able to proliferate at a great rate, but they also must be able to grow more slowly and then cease growing as the components of the developing child reach biologically determined maturity. When cellular control over the signal for "stop" is disrupted, the characteristic nonstop growth of cancers can occur.

The cells that make up organs are constantly exposed to the influences of whatever is carried through the body by blood and lymph. Blood itself is simply a collection of cells and their by-products transported in a watery medium.

The blood system contains three separate elements which, as Hippocrates observed in the fourth century B.C., settle into distinct layers in a test tube. On the top floats the transparent, wheat-colored liquid plasma. As well as transporting the minerals, sugars, salts, and other materials needed to nourish cells, plasma carries globulins, which are simple proteins that confer immunity against disease.

The deep bottom layer in the test tube is composed of red blood cells, whose sole function is to transport oxygen to tissues and to remove their carbon dioxide waste. Because they have no nucleus, red cells can't reproduce themselves. New ones must constantly be made in the bone marrow; approximately 300 billion are destroyed and replaced every day.

During its 120-day life, a red cell makes about 75,000 round trips between the lungs and the body's tissues to pick up oxygen and release carbon dioxide. At the end of their functional life, except for the few that may temporarily pass across the eye as "floaters" —the white or black spots that we occasionally notice crossing the line of vision—the red cells return to the bone marrow. There they are disposed of by scavenging white blood cells called leukocytes.

Clearing the body of useless cells is a continuous, basic activity of the white cells, which settle in the thin layer dividing the plasma and the red cells in the test tube. AIDS has made us aware of the importance of white blood cells by showing that diminished numbers or crippled functions of even a single subset can precipitate a chain of events that leads to death from overwhelming infections or cancers.

The white cell that is the major target of human immunodeficiency virus (HIV) is called the T4, or helper T cell. This cell is one of an array of white cells collectively called lymphocytes. As do red cells, lymphocytes begin their life in the bone marrow and must undergo many steps before they are ready to function.

White cells that are to become B cells, whose major activity is the production of antibodies, are programmed in the bone marrow. The antibodies they are designed to produce come in five classes (called immunoglobulins), each of which is involved in a different aspect of immunity.

White cells destined to become T cells migrate to the tiny thymus gland to receive surface markers identifying what their functions will be. Arriving in the thymus with numerous molecules marking their surfaces, these cells are put through a process of inspection. The majority are discarded, while other cells are selected for attributes such as an ability to discriminate between self and other. Some of them receive markers designating them as T8 cells with either suppressor or killer functions; others are programmed to be T4 helper cells. Once released into the blood, T cells spend their lives quietly resting in tissues such as the lymph nodes or patrolling the blood network for entities they recognize as foreign and therefore dangerous.

Dr. Irving Weissman of Stanford University, a cocreator of a mouse with a human immune system, has found that as well as being programmed with molecules that determine their functions, both B and T lymphocytes receive "homing" receptors that direct the cells both to recognize and to assemble in the vicinity of specific organs.

B and T lymphocytes cooperate in the immune response, and have both specific and nonspecific activities. The nonspecific lymphocytes include phagocytic cells (*phage* means "eater" in Greek, which is the literal function of several types of cells), helped along by a number of proteins called complement, which participate in causing the important inflammatory response against microbes. Triggered by various pathogens such as bacteria and viruses, the twenty interacting proteins of the complement system have many immune functions; among them is the ability to recruit other immune system components to infected or damaged tissues. Complement can also penetrate an invader's cell membrane, causing it to rupture.

Phage cells, of which there are several kinds, are the first major cells encountered by a pathogen. In response to signals sent out by inflammatory cells, phages move into these tissues, where they engulf and digest microbes or anything else foreign enough to attract their attention. Clotted blood, bacteria, damaged tissues, and dying cells also are eliminated by phages.

When activated by inflammation or the presence of an *antigen*, phage cells become mobile by repeatedly extending their "foot" or pseudopod and then pulling themselves after it. Under the electron microscope, phages can be seen creeping about the interior walls of blood vessels or squeezing through vessel walls to enter other tissues. When confronted by a foreign entity or damaged cell, the pseudopod reaches out to surround it. Once it is enveloped, the foreigner is bathed in the toxic enzymes contained within the phage cell, which completely destroy or break the stranger into bits. Filled, dying phage cells create the pus that is produced by a wound.

Depending on the type of target they engulf, some phage cells "decorate" their own surfaces with the strangers' molecules and present this new configuration to T cells, which recognize invaders only after their envelope markers have been manipulated in this manner by phages. The target may be a virus or any infected cell that displays part of a virus's identifying code.

Phage cells were first seen by a genuinely mad Russian zoologist, Élie Metchnikoff. Studying the transparent larvae of starfish under a microscope, Metchnikoff saw cells moving slowly toward and surrounding a rose thorn he'd stuck in the small creature. This activity was thought by the Russian to be the basis of immunity.

But the multifaceted competencies of the immune system, discovered separately by numerous researchers, pitted one theory against the other in the late 1800s. While Metchnikoff was deciding that phages are the basis of immunity, the American George Nuttall and the German scientist Emil von Behring observed that blood itself can kill bacteria. Nuttall insisted that Metchnikoff's phages were merely cleaning up bacteria killed by some substance in the blood. It required the balanced view of English physician Almroth Wright in 1903 to show that the two apparently opposing views were both correct. Blood contains the antibodies, and the phages are an essential element of the T-cell defense arm of immunity.

It has been learned since then that phage cells are a major part of the nonspecific immune response and are relatively indiscriminate in going after a range of biological debris. T and B lymphocytes, by contrast, have highly specialized functions.

T and B Cells and the Language of Immunity

B cells and T cells are the basic constituents of what are called the two arms of the immune system. The B cells and their products make up the humoral (blood) arm, and the T cells and their various chemicals constitute the cellular (tissue) arm. But all immune cell activities, whether general or specific, cellular or humoral, function through the interactions that occur between cell and target, between cell and cell, and within the cells themselves. These interactions occur in the form of molecules, the sum of whose activities constitutes the immune system.

Nobel laureate Niels K. Jerne likens the jumble of molecules that constitute the immune response to a vocabulary composed of "sentences that are capable of responding to any sentence expressed by the multitude of antigens which the immune system may encounter."*

The English language is generated from an alphabet of twenty-six letters; that of immunity, from the twenty-three chromosomes of the human genome. Depending on the precision required to get information across, it would be possible to substitute or to rearrange many of the words on this page without losing the sense of what is

*Niels K. Jerne, "The Generative Grammar of the Immune System," *Science*, Vol. 229 (September 13, 1985).

written. For instance, a phrase commonly found in reports of research is "much remains to be learned." "Learned" sometimes reads "done" or "elucidated." Alternatively, the phrase could read "not much more is known" or, inelegantly, "that's all we know now."

The "language" of immunity—which I'll call Imm—has to be as precise as is the selection of words for a Haiku poem or the instructions a doctor writes on the chart of a critically ill patient. To receive an Imm message, cells have to have the proper identifying markers; these markers are formed by the molecules called receptors that mark their surfaces. Markers on T cells have been supplied by the thymus; B-cell markers are programmed in the bone marrow.

Although it was once assumed that certain groups of cells are segregated from one another because of the great differences in their types and functions, it's becoming clear that whispered conversations are constantly being conducted between strangers, for example, between cells of the gut and those of the brain.

Many of the proteins that constitute the molecular language, such as the body's own antibiotics, the interferons, and insulin and various blood growth factors, can now be produced in the laboratory using recombinant DNA techniques that separate out a specific gene and grow it in quantities great enough to produce its wanted chemical messenger.

Our ability to perform such elaborate manipulations of cells and their products is a testimony to the explosion of knowledge about the immune system during the past few years. Not until the 1950s did researchers learn that only B cells produce antibodies, "the final products of an immune response that evolved to protect vertebrates from an environment filled with a seemingly infinite number of life-threatening infectious and toxic agents," a research group wrote. The existence of T cells was known, but not their important subsets or the range of their activities. The first important exploration of one of their major activities was done 100 years ago, when the first hint of the existence of antibodies was explored in Europe.

Antigens and Antibodies

In 1890, German scientists discovered two molecules that neutralized the poisonous toxins—the cause of clinical symptoms—produced by the diphtheria and tetanus agents. The hope that the molecules represented a single magic substance that could fight any infection was quickly dashed, however, because the molecule that handled one poison had no effect on the other. Each was a specific antibody designed to neutralize a different toxin.

A comparable scenario was proposed when interferons were discovered in 1957. For a while it seemed that these substances would provide a quick cure for cancers. It was subsequently discovered that although interferons do play a role in destroying cancerous cells, their effects are intimately linked with a large number of other chemicals and cells that also participate in controlling aberrant populations of cells. More specifically, in the body, interferons appear to protect only the cells in their immediate vicinity from various insults. Consequently, injecting interferons into the circulating blood causes extreme reactions because the interferons are not intended for systemic (systemwide) use.

Antibodies are produced each time we are infected by a novel virus or come across a poisonous plant for the first time. The substances that provoke this response are called antigens, which denotes something (anything from a virus to a molecule of the self) that elicits an immune response. Plant substances in particular, because they are so unlike animal tissues, tend to be powerful antigens: They are often used by doctors to test the strength of an immune response. Other, more specific antigens, such as dust or a suspect food, may be injected under the skin in minute quantities to determine what is causing allergies.

But an antigen isn't just a bad guy. A vaccination is an antigen that generates the creation of memory cells that can immediately react to the presence of diphtheria or tetanus, polio or measles, depending on which one the immunization is designed to work against. In the same way that letters of the alphabet can be manipulated to form an almost infinite variety of words, Imm is forever sorting and arranging its "letters" to deal with antigens.

Many of the markers on our cells can act as antigens, but the healthy immune system is programmed to tolerate them. When tolerance is broken or, as Dr. Weissman has recently shown, when immune cells escape the thymus before being "taught" to discriminate self from other, autoimmune disease is likely to follow.

Each B cell has a receptor that can recognize one particular antigen. In the presence of that antigen, the B cell becomes activated and churns out many millions of plasma cells, each one of which in turn makes millions of antibody molecules identical to the one programmed by the original B-cell receptor.

The five types of antibodies are called immunoglobulins, and each has a different role. These are designated Ig. IgG circulates in the blood and enters tissues; IgM primarily kills bacteria in the blood; IgA is found in body fluids such as tears and saliva and in the respiratory and gastrointestinal tract to protect the body's entrances from infection; IgD is a relatively mysterious molecule that is thought to participate in regulatory functions; IgE participates in allergic reactions.

An antibody is a new configuration of molecules and is therefore also "foreign," and antibodies called anti-idiotype antibody are produced against it. This anti-idiotype antibody likewise produces a response by the immune system, which makes another antibody, bringing to mind the cartoon of a large fish swallowing a smaller fish which has just swallowed a smaller fish, and so on into an infinity of infinitesimal reactions. These diminishing reactions serve to down-regulate the immune response when it is no longer needed.

Genetic Fingerprints

The immune system can't know in advance what it will encounter and therefore must be armed with a vast repertoire of recognition capabilities in the form of proteins. However, this is not a mechanical entity that runs along in a fixed way; underlying this enormous diversity is the genetic makeup of the individual. Virtually every cell in the body carries the several classes of genetic markers that are specific to the individual. Consequently, to a greater or lesser degree, the immune response is controlled by each person's genetic code.

Before an immune response can be generated and before antibody formation by B cells can commence, both the accessory cells (such as macrophages) and the T cells have to perform specific functions. Simply put, the accessory cell cuts a small section of the antigen—the strange molecules on the surface of a virus, for instance—which it displays on its surface. The newly marked cell then presents this antigen to a T cell.

The outer surface of the T cell is supplied with a potential receptor for this (or another) snippet of antigen. Because the T4, or helper, cell is responsible for initiating a number of chemical signals to regulate the immune response, its recognition of the stranger is a critical event. However, before this can happen, the portion of antigen has to be combined with a molecule supplied by the cell; this is what is called a major histocompatibility complex (MHC); this is produced by the genes of the individual.

These two molecules—the bit of antigen and that of the cell—combine to produce a new shape, a "fingerprint" that is recognized as strange, that is, as something that must be dealt with.

This interaction is essential. When the fingerprint created by the self antigen (the MHC) and the antigen of the germ fit together well, a series of appropriate events can commence. On the other hand, if the fingerprint is either too different from or too similar to the molecules of the cell, the immune system may fail to notice the invader and tolerate its presence instead of attacking it.

Alternatively, if the fingerprint is too similar to the molecular pattern of one of the body's own tissues, such as red cells or cells of the kidneys, when the destruct signal is turned on, those tissues are marked for attack.This in turn results in a form of disease appropriately called *autoimmunity*, such as diabetes or one of the arthritis conditions. This need for self molecules to participate in the activities of the immune system, along with the differences in individual genetic programs, surely accounts for many of the individual responses to infectious agents that are routinely described in the following chapters.

When the specific genetic susceptibilities of individuals to infections, cancers, and autoimmune conditions are finally identified, vaccines and individually tailored treatments to correct or prevent these conditions will revolutionize the practice of preventive/curative medicine. The ability to add or remove receptors that identify cancer cells, for instance, will turn the immune attack against these aberrant cells. Alternatively, autoimmune conditions will be circumvented by hiding or getting rid of antigens that mark healthy tissues for attack. While all this is in the future, it is a near future, given the rapid expansion of knowledge in this field.

Immune Against the Self

Our individual genetic patterns clearly predispose us to be harmed by or protected from a multitude of external influences. These influences include stress, polluted air, smoking, food and allergens, and susceptibility to cancers and infective agents.

In many ways we are like the houseful of unique items we accumulate over the years through inheritance or because we happen to like or need them. Some are good, some bad; some work, others don't. In the context of health, the content of our individual "houses" depends on factors in random combination, such as the genetic markers carried down the generations in heritable germ cells, sex, age, and molecular variables too chaotic to consider given the great gaps in present knowledge.

In each of our hypothetical houses there is a locked case containing a collection of loaded guns, which lack only triggers to begin firing. If and when individuals who are either genetically so programmed or just plain unlucky are confronted with antigens that represent the triggers, the guns begin firing at targets within the self. That's autoimmunity, a process that can affect virtually any tissue or organ.

Immune "Dyslexia"

Dr. David Katz describes autoimmunity in a generic way by referring to it as dyslexia of the immune system. Dyslexia, which is a disturbance of the ability to read, is comparable to dysfunctions that cause the loss of communication between various cells of the immune system. When this occurs, the immune system is no longer properly regulated, and tolerance of self antigens is lost. Any number of malfunctions may ensue.

Discussing the "incredible parallel" between AIDS and well-known autoimmune diseases—such as thrombocytopenia purpura, which causes small blood vessels to become fragile and frequently leads to inflammatory arthritis—Katz offers a series of mechanisms that probably play a major role in the destruction of immunity that characterizes this syndrome. But while the outcome of damage to essential immune cells is most obvious in AIDS, similar deficits doubtless underlie all autoimmune phenomena.

Katz, head of the Medical Biology Institute in La Jolla, California, proposes that both antibody (B cell) and cell-mediated (T cell)
mechanisms theoretically could account for the multitude of dysfunctions in HIV-infected persons but inclines toward the possibility
that "one or more HIV retroviral gene products actually become
integrated in the cell membrane."

If this is what happens, T cells that should be dealing with the
virus are themselves targeted for destruction because they carry the
viral antigen embedded in their envelopes. This is probably the way
in which a relatively noncytopathic virus—a virus that is a poor
killer of cells—manages eventually to virtually eliminate the T4 cell
population.

Another possible, or perhaps additive, mechanism is that of molecular mimicry. Since the enzymes and hormones that perform
essential biological functions tend to be conserved in all species and
passed down over evolutionary time, the host and the invading
pathogen may have short segments of similar or identical molecules.
Some pathogens synthesize antigens so similar to the cells of the
host that they are therefore ignored. When the similarity is this
great, it is referred to as *molecular mimicry*.

Dr. Michael Oldstone, an original creator of the molecular mimicry paradigm, writes, "Conceptually, molecular mimicry can produce autoimmunity when virus and host determinants are
sufficiently similar [molecule sequences that are the same] to induce
a cross-reactive response yet different enough to break immunologic
tolerance [of the self]." In other words, the immune system is "dyslexic" in that it has no way to distinguish between a dangerous agent
and its own tissues, which are identified by markers that will now
elicit an attack.

Oldstone claims that antibodies against hormones, white cells,
and cells of the nervous, immune, and endocrine systems as well as
the gut, heart, and muscle are generated by a panoply of viral agents.
Coxsackie B virus, for instance, is sometimes found in patients with
myocarditis, an inflammatory autoimmune heart condition. In this
case the similarity between the antigen's fingerprint and components
of the heart muscle may result in both being marked for attack.
The same virus has recently been found in the muscle tissues of
people suffering from chronic fatigue syndrome (Chapter 18). Generally speaking, a continual immune system attack creates inflam-

mation, which heightens and continues the immune response. This usually accompanies and severely contributes to the damage caused by chronic autoimmune disorders.

When a genetically susceptible person is infected with a virus whose molecules mimic the Imm code of normal cells, the antibodies developed to fight the invader also destroy normal cells. Meanwhile, the virus has been eliminated, so that the immune attack is no longer needed, but a chronic process has been set in motion in which the normal tissues remain under constant attack. Many rheumatologists think this may happen in some cases of rheumatoid arthritis.

Rheumatologist George Ehrlich, while practicing in Philadelphia in 1981, stated that he knew but couldn't prove that he had managed to abort hundreds of cases of rheumatoid arthritis by vigorously treating the first signs of the disease with heavy doses of anti-inflammatory drugs. Ehrlich is convinced that he has prevented what could have developed into the self-perpetuating cascade of joint destruction that characterizes this disease.

Katz cautiously proposes a similar approach to early HIV infection: By utilizing one of the drugs that suppress the immune system enough to prevent rejection of transplanted organs, one might be able to abort a comparable cascade of immune system dysfunction.

Many of the disease conditions caused by the viruses we will be discussing in the following chapters have autoimmune components. Most are transitory and, although bothersome and often painful, disappear when the infection that has kicked them off ends. But an unfortunate minority of people suffering from viral infections —for many reasons, some known and others still mysterious—experience long-lasting, sometimes deadly consequences.

Although these types of autoimmune disorders seem far removed from what we think of as the outcome of viral infections, researchers have proposed mechanisms by which they may occur. These include molecular mimicry; the virus making the infected cell or some component of it appear foreign to the host's immune system; the virus triggering autoimmunity by overstimulating or destroying critical cell populations (as in AIDS); or a virus triggering antibodies against newly formed antibodies—the big fish eating the smaller one eating the still smaller one, etc.

California researcher Raphael Stricker suggests that the nerve damage and loss of sensation in the extremities so common in people with AIDS may result from a virus-induced attack on the nerve cells. But Stricker, who has identified autoantibodies against nerve cells, reports that these are only one of the more than twenty autoantibodies identified in these patients.

"These are a sure sign that the immune system has gone haywire," Stricker maintains, adding that "this supports the possibility that AIDS is primarily an autoimmune disease." Many of these symptoms are similar to those caused by multiple sclerosis; others resemble severe arthritislike disorders, such as Reiter's syndrome and Sjögren's syndrome (Chapter 15).

Another aspect of autoimmunity is that conglomerates of antigens and antibodies called immune complexes are often formed. In diseases such as systemic lupus erythematosus (lupus), these relatively large immune complexes clog the small tubules of the kidneys, eventually producing severe problems.

Since many of these disorders can be temporarily relieved by "washing" the blood by running it through a machine that removes autoantibodies, these autoantibodies, rather than the virus itself, are apparently responsible for the damage. Autoimmunity thus represents a condition in which an individual's immune system has circumvented its natural lack of response to self antigens. The presence of a virus whose antigens resemble those of the self may upset the balance and establish destructive cascades even after the virus is gone.

In summary, we have seen how important the idiosyncratic, genetically determined response of the individual is in dealing with infective agents. The partially understood contribution of the host is central to the outcome of an infection. But each pathogen has its own special agenda as well, which the following chapters on specific diseases will describe.

PART THREE

THE DISEASES

Introduction

The organization of the following chapters has been difficult because as one textbook puts it, a single virus can arbitrarily be classified "simultaneously as a poliovirus, an enterovirus, a neurotropic virus and a picornavirus."* In other words, the poliovirus infects through cells of the gastrointestinal system, from where it may travel to the central nervous system. Even for a virus, it is extremely small, hence the designation *pico*.

But the picornavirus family includes the hundred or so viruses that cause the common cold as well as the coxsackievirus, which can infect the mouth, upper respiratory tract, and stomach, and so organization by body systems doesn't work. Nor does severity of symptoms. The herpes simplex virus that causes fever blisters sometimes enters the brain, where it may cause death or affect emotional and intellectual functions.

"Dropping" a virus into a body in many cases is similar to mixing one color of paint into another. Depending on whether the pail is jiggled or whether the new color is stirred in vigorously, diffusion throughout the old color takes many forms. Sometimes added paint

*G. Youmans, P. Paterson, and H. Sommers, eds. *The Biological and Clinical Basis of Infectious Diseases*, 2nd ed. (Philadelphia: Saunders, 1980).

seems to sink rapidly and, except for some slight pattern left where it was poured, apparently disappears.

By the same analogy, a virus may take an infinite number of "shapes" when a person is infected. Depending on whether and how the mixture is stirred, swirls of color may go in many directions or, in the case of a virus, may affect numerous body systems.

With that in mind, most of the following chapters focus on the location of the cells that are targeted by viruses. For example, polio, although relatively inconsequential as a cause of stomach disorders, is included in the chapter about gastrointestinal viruses (Chapter 11) because of its initial affinity for cells of the stomach.

The chapter on influenza (Chapter 9) also includes viruses that apparently drive some coronary disorders because there is an association between the influenza virus and degenerated heart muscles. By the same token, multiple sclerosis (MS) isn't included in the chapter on viruses that are known to affect the brain and central nervous system (Chapter 17) because MS is primarily caused by an autoimmune response to virus.

A new virus, HHV-6, has recently been shown to cause German measles (rubella), a mild, unimportant childhood disease but one that, when contracted in the first three months of pregnancy, is known to cause severe malformation and other dire effects on about half the affected fetuses. But HHV-6 may also be implicated in what is called chronic fatigue syndrome; it may also play a role in Hodgkin's lymphoma, leukemias, and AIDS. But excepting the hundred or so that cause "colds," the often arbitrary "stirring" pattern outcome applies to most viruses.

5

Virus and Cancer

In recent years the American public has been bombarded with stern admonitions not to smoke or to tolerate anyone who does and to eat or avoid foods with a particular consistency, coloring, preservative, or pesticide. There are dire warnings that sexual activity spreads viruses (aside from human immunodeficiency virus) which cause cancer. We are told that there's no doubt that many of these habits and substances *cause* cancers. Some probably do in a roundabout way in people who are particularly susceptible because of underlying genetic makeup plus combinations of nutritional and environmental factors, but there is virtually no knowledge of what these factors may be in an individual, much less in populations.

Nevertheless, much of the public has been swept up by a militant Calvinist spirit of sacrifice with moralistic overtones in the hopeful assumption that by following current medical dictums, it won't be subject to the most feared diseases. Some of these sacrifices may be beneficial to the individual; others may not. But despite the complexities of the genesis of most diseases, stern and simplistic avoidance messages are being cranked out with dazzling frequency.

Significantly, many of these public health messages concern the triad of large-scale killers that physicians have little success in treating: AIDS, heart disease, and cancer. The ball has been tossed into the patients' court: Some doctors have even debated whether they should have to take care of patients who don't follow current medical dictums or whether taxpayers should have to bear the cost of treating conditions that we "bring on ourselves."

But Marshall Becker, professor of public health at the University of Michigan, has written in the *Lancet* that "the domain of personal health over which the individual has direct control is *very* small when compared with heredity, culture, environment and chance."

The scientific basis for most prevention messages is at best flimsy and at worst irresponsible. A woman with breast or cervical cancer who hasn't had a yearly mammogram or Pap test is automatically guilty of negligence, though mammograms might or might not have revealed the new growth and the Pap might or might not have been accurate. An article in the *Journal of the American Medical Association* in 1989 by Dr. Leopold Koss described the Pap test as "a triumph and a tragedy," because despite the curability of cervical cancer when it is caught early, "the accuracy and interpretation of the test itself is often inadequate."

Although many victims of one type of lung cancer (the other, small cell cancer, seems to depend on a miscoding gene) have smoked, the majority of smokers never get cancer. According to studies in various medical journals, female smokers appear to be protected against rheumatoid arthritis and osteoarthritis and seem to be less prone to ulcerative colitis. This kind of conflicting evidence clearly shows that very little is known about the necessary combination of factors—genetic and environmental in the form of carcinogenic influences such as viruses, radon, pollution, and alcohol consumption—that play a role in determining whether one will develop cancer in association with tobacco smoking.

Sex, outside the narrow parameters of monogamous relationships, has always been a morality-loaded catchall for disease as the wages of sin. For instance, the idea that prostitutes have more cancers of the cervix than nuns do is repeatedly raised without reasonable evidence. Since the papillomavirus is likely to play an important role in cervical cancers in women, it makes sense that the more partners, the greater the likelihood of becoming infected with this virus. However, China, with an extremely low rate of venereal disease (a marker for sexual activity), has one of the highest rates of cervical cancer in the world.

The point is that many of the specific and most of the vague recommendations are based on small and not always well-conducted studies of special populations. The variables associated with un-known factors in most cases are too great to translate what may be the case in Eskimos or rural Chinese to the entire American pop-ulation.

Dr. Dimitri Trichopoulous of Harvard cites the currently pop-ular idea that keeping her weight down will prevent a woman from developing breast cancer. Not necessarily so, says the epidemiologist, because the influence of estrogen on a woman begins before birth and because breast cancer may depend more on the actual amount of mammary tissue (which is related to the size of the woman) than on the plumpness conferred by fat.

People who avoid meat and cheese may be depriving themselves of the conjugated linoleic acid they contain, which has recently been shown to be a powerful anticancer substance. No one has recom-mended stoking up on cheeseburgers to prevent cancer or smoking as a defense against arthritis, but these and many other contradic-tions show how difficult it is to make simple pronouncements about diet and disease, particularly in regard to cancer.

Human-made carcinogenic substances are probably a minor fac-tor compared with those found routinely in nature, such as sunlight, viruses, radon in the soil, and food and plant poisons. But the public feels a false sense of security when a substance such as Alar (dam-inozide), which is used to ripen red apples, is removed from the market or when their diets are lopsidedly composed of roughage to prevent colon and stomach cancer.

Some day soon it will be possible to examine individual genetic patterns and say for sure that because a chromosome lacks a certain cluster of amino acids, an individual is at high risk for specific conditions such as breast cancer, arthritis, Alzheimer's disease, and even certain infections. At present, however, microfiddling with the habits of the whole population over such niceties as eating broccoli without hollandaise is a far cry from science. Despite this, the mes-sage is clear: Changing your habits will prevent these conditions. Maybe so. Maybe not.

But one thing is sure: Living in a state that Dr. Petr Skrabanek of Dublin calls "the holy dread of preventionism," in which a vague sense of dis-ease accompanies doing anything that's fun, fat, or fool-ish, smacks more of medicosocial control than good health advice.

For the majority of people who will or won't develop one of the major killer diseases before achieving a reasonable life span, luck, good or bad, seems to depend on the degree of susceptibility or resistance to disease, on a kind of molecular fine-tuning of their inherited genetic programs.

Even the AIDS virus has very different effects on different individuals. During the early days of the epidemic it appeared that anyone exposed to the virus became infected, but time has shown that some people remain uninfected despite repeated exposure. While older people succumb quickly, those infected as adolescents have been found to live longer and in better health. Why this is so remains a mystery. It is also probable that some people have been infected and gotten rid of the virus without ever knowing they'd been exposed.

Another unanswered question about AIDS concerns Kaposi's sarcoma, which may not be a true cancer. Early in the epidemic, the great majority of men suffering from this condition had the same special genetic marker as did the men who classically contract slow-moving Kaposi's sarcomas in their sixties. As more people became infected with the human immunodeficiency virus (HIV), those without this special genetic marker have also developed Kaposi's sarcoma, but the disease is still almost exclusively found in gay men with HIV infection. Few women or intravenous drug abusers develop Kaposi's sarcomas. Why? Subtle differences in hormones, coinfections with other agents, lifestyle? No one knows.

Many but not all AIDS patients develop unusual cancers, such as those of the lymph system and brain. HIV itself doesn't directly causes cancers, but it does participate in a multistep process in which the virus knocks out immune cells that should be monitoring the system for potentially cancerous cells. To explain how cells become cancerous—a process that essentially involves the cells' loss of the signal to cease growing at an appropriate stage—one would have to understand what happens to the cells' genes as well as the regulation of the sugars, proteins, carbohydrates, and enzymes that regulate their activities.

Only a tiny fraction of a cell's genes are active at any given time, and almost nothing is known about the signals that turn them on and off. It is usually impossible, therefore, to cite a simple cause-and-effect relationship between what we do and what eventually ails us.

How a normal cell turns into one that produces a cancer can better be understood by analogy. A woman with a broken ankle is brought from a football stadium to the hospital emergency room, a straightforward case of trauma. But if it is necessary to identify the actual underlying cause of the fracture, the picture becomes very complicated. Someone in the crowd pulled a flask from his pocket, had one drink too many, and became verbally abusive toward those around him as the fans began to leave the arena. A fight ensued in which the woman was knocked down and her ankle was broken. Is the "cause" of her injury the personality of the troublemaker she never saw, the drinks he took, or her having chosen to leave the stadium through a particular exit?

Perhaps she didn't get much sleep the night before and her reflexes were slow. We might take the preventionist way out and decide that large gatherings must be prohibited to prevent such injuries. Or we might follow the scientific path to discover what those investigating the genesis of cancer are discovering: that many elements feed into the cascade of events, some random, others given, leading to the development of the uncontrolled cell growth that is cancer. One cancer-causing input over which we have little or no control is the sequence of events that follows many viral infections.

Cancers, Viruses, and Oncogenes

Joshua Lederberg raises many fascinating questions when he writes, in the *New England Journal of Medicine*, that

> From the perspective of the virus, the ideal would be a nearly symptomless infection in which the host is oblivious to providing shelter and nourishment for the indefinite propagation of the virus' genes. Our own genome carries hundreds or thousands of stowaways. The boundary between them and "normal" genes is quite blurred. Not much more than 1% of our DNA can be assigned specific physiological functions: most of it is assumed to be a "fossil" legacy of our prior evolutionary history, DNA that is today parasitic on the cell.
>
> Further, we know that many viruses can acquire genetic information from their hosts, which from time to time they may transfer

to new ones. Hence, intrinsic to our own ancestry and nature are not only Adam and Eve, but any number of invisible germs that have crept into our chromosomes. Some confer incidental and mutual benefit. Others of these symbiotic viruses or "plasmids" have reemerged as oncogenes, with the potential to mutate to a state that we recognize as the dysregulated cell growth of a cancer.

This is a form of Darwinian evolution that momentarily enhances the fitness of a cell clone at the expense of the entire organism. Still other segments of "nonfunctional" DNA are available as reserves of genetic potential for further evolution, in a sense more constructive for the individual and the species.

In short, what Dr. Lederberg means is that a lot of unknown information is contained in each of our genes and that much of it is involved with viral infections, both ancient and contemporary.

It won't be many years before we have a firm fix on many of the diseases that afflict humans, because the United States in 1988 began a concerted effort to map the entire human genome. With Dr. James Watson of DNA fame at its helm, the project—similar in terms of the energy being devoted to it to the Manhattan Project—will eventually identify all the genes in the human body.

Thomas Caskey, director of molecular genetics at Baylor University, likens his team's part in the project to "traveling the human genome as if it was a coast-to-coast highway," along which "cities" (genes) are placed here and there to zero in on the 400 human diseases known to reside on the X (female) chromosome. The team's high-tech diagnostic technique has been applied to three heritable diseases.

The Baylor team has used a virus to replace a missing gene whose absence causes adenosine deaminase deficiency, the most virulent form of severe combined immune deficiency syndrome (SCIDS), a congenital condition resembling AIDS. The replaced gene then begins faithfully directing a corrected message in mice as well as in human bone marrow cells. The replacement of this missing gene in affected children has recently been approved by federal authorities.

When and if they can be identified, most of the "time bombs" represented by missing or incorrectly located genes will be correctable. In other words, if the body's brakes and accelerator have been knocked out by a genetic mistake at the top of a steep hill, genetic engineers will be able to reline and adjust them. To do so, of course, requires that the exact location of each gene be pinpointed.

One of the problems inherent in the genome project, as Lederberg explains, is that the majority of genes have no function: They are merely fossils that have been left behind from millennia of viral infections and passed on from one generation to the next. This happens because a virus has to install part of its own genetic program in order to infect a cell. If these viral gene sequences are incorporated into the cellular genetic pattern, they can be transmitted along with functional genes.

Most of what is known today about the human genome has come from studying diseases. For instance, by looking at the DNA in cancers and tracking backward to see which normal gene it correlates with, scientists have been able to identify about fifty oncogenes. Undisturbed, oncogenes are responsible for regulating the growth and division of cells; disrupted by any one of a number of changes, including infection, these genes play a major part in the development of cancer.

Oncogenes

Michael Bishop of the University of California in San Francisco, who with Harold Varmus was awarded the 1989 Nobel prize in physiology for oncogene research, has long been interested in how damage to DNA is involved in the genesis of cancer. Bishop writes that the expression of a previously regulated or silent gene can arise when a virus infects one cell, picks up from it a normal gene, and carries the normal gene to another cell. The gene from the first cell may then be inserted in the wrong place in the newly infected cell's DNA, causing that cell to lose its usual control over how and when it grows.*

When oncogenes were first identified in the Rous chicken sarcoma virus in the mid-1970s, it initially seemed that the virus's ability to cause cancer was merely a side effect of its own need to replicate and that one of the chicken virus's own genes serves as the promotor of cancer. But when one of the virus's four genes (called *src*) was

*Michael Bishop, "Oncogenes," *Scientific American* (September 1984).

cut away by special enzymes and the remainder of the virus was injected into other birds, it failed to cause cancer in chickens. The *src* gene (named after the sarcoma) alone was able to cause the malignant transformation of cells. Without *src*, the virus could reproduce itself, so the gene appears to be irrelevant to the virus's life cycle.

What, then, is its origin and function? Why, with its limited repertoire of genes, would a virus have one that is unneeded in its own life cycle, a gene capable of killing the host cell on which the virus depends? Bishop and his colleagues set about to investigate a proposition raised by National Cancer Institute researchers: Retroviruses such as Rous carry oncogenes and in effect introduce them into the cells they infect.

But as more detailed descriptions were made of the way the *src* gene is assembled, it emerged that this oncogene, as well as those from some other viruses, is made up of strings of amino acids that are picked up from the *host's* cells rather than being part of the virus's own genetic equipment. Thus *src* and other oncogenes have been integrated into viruses from infected cells and carried about as excess baggage.

Viral and cellular *src* genes produce proteins that are identical. As Bishop describes it, "It is as if the two proteins were designed for the same purpose, even though one is a viral protein that causes cancer and the other is a protein of normal cells."

When the Rous sarcoma virus infects a cell and incorporates its genetic material into that cell, the *src* gene from a long-forgotten host cell is transcribed along with the viral genes. Normal regulation of the cell, which is now controlled by the virus-provided oncogene as well as by the cell's own genes, is lost, and the sarcoma develops quickly.

This is a relatively clear-cut example of the effects of a single gene carried by a virus as a promoter of cancer. Several other scenarios are thought to be possible. One is that an individual may be born with a genetic pattern in which one or more genes designed to control aberrant cell growth are missing. All may go well until a virus or another cancer-promoting event causes changes in the cell's growth signals.

An example of this is the gene called *rb*, which was originally found in children with retinoblastoma of the eye but subsequently was discovered to be present in all human cells. (Note that this gene, named after a cancer, and oncogenes are both normal constituents of the genetic pattern. However, oncogenes and the retinoblastoma gene were given their names when they were thought to participate only in malignancies.) If the *rb* gene is missing or is formed in an irregular way, this encourages the development of several cancers.

When this rare eye disease is diagnosed and treated early in children, the condition can usually be cured, but these children have a high subsequent risk of developing other neoplasms (*neo* means "new," and *plasm* means "growth") later in life, particularly bone cancer. This indicates that people with this single genetic defect live under the lifelong influence of a process that puts them at high risk for cancer. Unlike some other cancers, in which extra or out-of-place genes seem to account for uncontrolled cell growth, it's the absence of the *rb* gene that allows the development of neoplasia.

Abnormalities of the *rb* gene are found in many human tumor cells, including those taken from cancers of the breast, bladder, and bone and from small cell lung carcinomas. In many but not all cases, the loss of this gene or changes that render it unable to produce a protein called p105-*rb* correlate with increased cell growth. It seems that p105-*rb* functions as a chemical off switch. In this instance, the predisposition to cancer already exists in the genetic makeup of the individual, allowing other influences to eventuate in cancers.

If a virus itself carries an oncogene or if its own genetic material integrates into a cell in such a way that it causes the cell's own controls to become scrambled, the potential for cancer is also created. This potential exists already in the normal cell, whose own assembly of genes includes proto-oncogenes.

In short, the development of some cancers appears to result from a cooperative venture between viral and cellular genes, plus (as was shown by Ludwik Gross) factors related to the age at infection and, of course, the ability of the immune system to recognize and control these outlaws when they arise.

Alternatively or in an additive fashion, a dose of virus or radiation may be so great that it overwhelms a normally functioning regulatory system. It's probable that cancer development usually depends on several oncogenes being activated in sequence. Researchers have recently found that in the case of small cell lung cancer, as many as fourteen changes must take place to cause the transformation to malignancy.

While a few viruses such as Rous have been shown to cause cancer in animals, because such experiments can't be carried out in people, this is far more difficult to prove in humans. But something happens in individuals that allows even the most ordinary, widespread viruses to drive the progression to cancer, even though other, unknown factors are involved. The normal chemicals that serve as growth factors are clearly implicated. Cells stimulated by them are strikingly similar to those changed by retrovirus infections (Chapter 6).

Altogether, viruses make poor guests. Bishop describes the relationship of a virus with its host cell as "depending on the lavish hospitality offered by the cell" while the virus retains much authority to control events. As small as they are and as limited as they are in their genetic information (a single cell has a genome containing tens of thousands of genes; viruses can have as few as four or five genes), viruses enter cells and take over the orchestration of their life cycle, using whatever they need to make more of themselves. Often the infection is innocuous and virus and cell go about their usual business undamaged and unchanged, but just as often this is not the case.

Papillomaviruses

In early 1989, Palmer Beardsley of the University of Texas said at a conference on tumor-causing DNA viruses that "we know that 20 percent of world cancers are virus-induced." The conference was specifically concerned with three ubiquitous viruses: hepatitis B (Chapter 12), Epstein-Barr (Chapter 8), and papilloma. The mechanisms by which hepatitis causes liver cancer appear to be different from those of the other two viruses, each of which has its own way of evading the immune system to initiate cancers. The papillomavirus apparently is able to alter the host's *rb* gene.

If you've had warts, you've been infected with one of more than fifty known papillomaviruses. Warts on exposed skin—hands, feet, etc.—are caused by several strains, while those in the genital area are caused by others. The most worrisome outcome of infection with this widespread virus is the suspicion by many scientists that papilloma plays a major role in the development of female cervical, and possibly other genital, cancers.

The unfolding saga of this common pathogen is a classic example of the recognition of a virus that participates in the genesis of cancer. Papillomavirus was identified in the 1930s but has not yet been grown in the laboratory, so little was known about it until recently, when modern techniques made it possible to grow the virus's DNA.

Identification of different strains of the papillomavirus have accumulated since the late 1970s, and today more than fifty-three are recognized, with more likely to be found. Types 1, 2, and 4 cause warts on the hands or the soles of feet; strains 6 and 7 produce warts in the larynx. Infected women may pass on strains 6 and 11 to their newborn infants, infrequently causing juvenile-onset respiratory papillomatosis, in which warts on the child's vocal chords must be removed, sometimes again and again.

Positive Pap tests often indicate the presence of dysplasia, a precancerous overgrowth of normal tissues caused by papilloma strain 16. Some studies show that nearly all women under age fifty are infected with papilloma 16, and in the United States about 7,000 die each year from cervical cancer. Evidence suggests that anal and penile cancers are associated with papilloma as well, and traces of the virus have been found in the semen of men with severe chronic warts and those whose female partners are infected.

The *Lancet* published the results of a recent study from central China, where rates of cervical cancer are among the highest in the world, the incidence of venereal disease was shown to be negligible among the ultraconservative farmers, yet 90 percent of the cervical tissues examined contained viral DNA. Given the "puritanical mores" of this population, the Chinese scientists questioned whether papilloma can be spread by casual contact as well as sexual contact.

Papilloma is also implicated in the skin cancers of persons who have had kidney transplants. To prevent the rejection of a transplanted organ, the immune system must be suppressed for the remainder of the recipient's life. In sixty-nine Scottish transplant patients who lived more than five years, the rate of various serious skin conditions was found to be extremely high, with 77 percent having viral warts, 19 percent having keratosis (a precancerous skin condition), and 12 percent having actual skin cancers. The genetic material from papillomaviruses 5 and 8 was found in virtually all these patients.

Until recently the proposition that papilloma causes cancer generated heated arguments pro and con, with those against citing the almost universal infection rate of the cervix (close to 1 million new cases are seen annually in the United States) and the relatively small numbers of cancers that develop. But the virus is found in almost 100 percent of cancerous cervical cells, and if the cancer spreads to other organs, the virus is also found in those distant sites, indicating far more than a casual relationship.

Peter Howley of the National Cancer Institute recently presented evidence that may lay these disagreements to rest: The virus seems to be able to shut off the regulatory proteins of the *rb* gene.

In the test tube, scientists have shown that proteins of papillomavirus type 16 (as well as simian virus 40 and the adenoviruses, which cause tumors in rodents) can block the activities of p105-*rb* —a regulatory gene product—which then may fail to control the growth of infected cells in much the same way that the protein is found to be blocked in the cancers mentioned earlier in this chapter.

However, since most adults have been infected with both papillomaviruses and adenoviruses and few develop associated cancers, it seems that an underlying genetic difference determines who will and who won't develop a cancer triggered by these viruses. It may take a combination of age at infection, several viruses working together, and dozens of mysterious events, such as whether cells are dividing, to cause cancer.

Whether cancer follows papilloma infection apparently depends on the ability of these viruses to become latent. An active infection kills cells and calls into play immune system elements that get rid of the virus: When a virus manages to remain hidden for long periods of time (i. e., becomes latent), the chance that it will interfere with the cells' own programming and thus create abnormal growth becomes a real threat.

But what allows ordinary virus to become latent in these relatively rare cases? Why do women who smoke appear to develop more cervical cancers, though fewer of the uterus? Levels of sex hormones, which are thought to be slightly lower in smokers, may play a role; the state of one's immune system, which becomes less efficient with age, must be important; coinfection with other viruses and long-standing inflammation (as occurs when the liver is infected with hepatitis B) probably come into play. Underlying everything are unknown genetic predispositions. The bottom line: Take pre-ventionism with a grain of salt—or something.

6

Retroviruses and Cancers

The Shaman

All societies have shamans, men and women who are thought to be able to intercede with the unknown world of the gods. While some shamans doubtless believe completely in their own powers, others, and perhaps all to some extent, recognize that the tricks they employ to convince the populace are indeed just that—tricks.

Anthropologists who have studied their belief systems relate that many shamans are afraid of their fellow practitioner down the road, assuming that while they have to resort to mumbo jumbo, the other shaman has the real stuff. Today, as religious beliefs appear less convincing than they once did and the technological world becomes progressively less accessible to common knowledge, scientists are often perceived by themselves and others as contemporary shamans, interpreters of the mysterious, keepers of the faith.

For the medical profession, this perception began in earnest when penicillin became available in the 1940s. At once physicians became the purveyors of a magic bullet against the infections that had killed so many of their patients. Instead of worrying at bedsides, doctors began writing prescriptions. Just as the discovery of anesthesia had pushed primitive and grisly surgical procedures into a new era less than 100 years earlier, by warding off the infections that too often followed long and complicated surgery, antibiotic use made possible today's advanced techniques.

During the years between World War II and the late 1960s it even seemed possible that cancer would soon yield to newly identified chemotherapies. Medical science in general was buzzing along nicely, with birth control pills, heart transplants, the discovery of cortisone, the development of vaccines against childhood diseases, and innumerable new categories of drugs to treat everything from tuberculosis to severe mental disorders. For a long time there was a kind of general euphoria over the improved practice of medicine. But along with the tide of rising public expectations and in the wake of the space program, another entity evolved that took science into a new dimension.

The creation of the media "science beat" was to have a remarkable effect on its subjects, particularly the medical field. It might have seemed unimportant at the time, but when reporters began actively raising questions about accepted treatments whose outcomes didn't seem very good, when schematic drawings of body systems and molecules became a regular feature in new science sections of print and electronic media, what the shamans of science had been doing behind high walls gradually began to come under public scrutiny.

Researchers became stars, decisions about funding began to be influenced by "names," and much of the mystery evaporated in the light of probing, science-wise reporters. The magicians were human, after all. Sometimes they were too human; during the past ten years, many of them have publicly fallen from grace when fudged or contrived data, fictional research, and outright fraud have been exposed by their peers and publicized. These episodes have shaken the confidence of the general public in the formerly awe-inspiring shamans of science.

Many of these former shamans succumbed to the horrendous pressures in academic and research circles to publish—publish or perish professionally, publish or forget funding and all it implies. The stockpiling of citations—some researchers' résumés have literally hundreds of pages of references to their published works— has been described as securing a reputation by repetition. Ample opportunity for this is provided by the growth of scientific journals, at least 40,000 of which are now cranked out across the world.

Recently, fraudulent work has been exposed in the areas of experimental psychology, heart and pediatric research, and anthropology, all done by young men who had been on the way to making a name for themselves by producing inordinately large numbers of articles for their respective professional journals. This in itself might have served notice of something being out of kilter: With all the writing, what time was left to do the work? Their peers finally noticed, however, and convened review boards which relieved the guilty of their posts and ended their futures in science.

A Difference in Style

The importance of publishing and the dubious peer review process by which many journals are suspected of selecting papers based on researchers' names and affiliations as well as what happens when one goes outside these sacred channels to the mass media were never better exemplified than during the search for the virus that causes AIDS. Events that were in part to shape the management of the emerging epidemic began innocuously enough, but they eventually affected the entire conduct of the research. Researchers who might have joined in earlier were put off, the paranoia of the groups that originally were found to have AIDS was exacerbated, and researchers were polarized so that important information often was withheld rather than shared.

An unusual gang of papers about a single issue—AIDS—appeared in the May 1983 issue of the widely read general science journal/magazine *Science*. There were five articles. Myron Essex from Harvard wrote that his group had found evidence that eighteen of seventy-five AIDS patients were infected with human T-cell leukemia virus (HTLV), the first viral family name assigned to the AIDS virus, suggesting "that at least 25 percent of AIDS patients have exposure to HTLV or a closely related virus." Robert Gallo's National Cancer Institute (NCI) team reported finding HTLV-infected white blood cells in two of thirty-three men with AIDS, and a second paper from the same group described the isolation of the leukemia-causing HTLV virus in an American and two French women with AIDS.

A fourth paper, from Luc Montagnier's group at the Pasteur

Institute in Paris, described the finding in an AIDS patient of a virus which the French suggested belonged to the HTLV family but was "different immunologically." "The role of this virus in the etiology of AIDS remains to be determined," the paper concluded.

However, the centerpiece of the unpleasantness to come was presaged in a line that read "All attempts to infect other cells such as B-lymphoblastoid cell line (Raji), immature or pre-T cell lines and normal fibroblasts were unsuccessful." In other words, they had a new virus but couldn't get it to grow.

Montagnier found a virus, identified it as HTLV-I$_{ep}$ (for the initials of the patient from whom the cells came), but couldn't keep it alive long enough to test its effects on cells. Gallo also had several similar viral isolates from AIDS patients and was searching for a way to infect cells without killing them. Otherwise, the quantity of material needed for critical analyses simply wasn't available.

"If I lived this whole event over again, what I would wish was that we had analyzed the French virus, which we had the power to do, in those first papers," Gallo says ruefully. The first time the virus was detected in the NCI lab was in February 1982, by which time "we had three or four virus types," he says.

As had been done many times before, plans were made for the NCI and the Pasteur Institute to collaborate on identifying the new viruses, with the Americans to do the molecular biology. But this cooperative venture never materialized, and Montagnier is said to blame its failure on what happened at a small elite science gathering at Cold Springs Harbor on Long Island a few months after the *Science* papers appeared. Gallo asked some very specific questions about the French research, and Montagnier was offended.

"I can't to this day really understand it," Gallo says. "I've had questions like that thrown at me a thousand times. You don't take them personally, these questions of science."

Gallo describes the hard feelings generated there as being "simply a difference of style."

"If it had been an English meeting, it would have been mild; for an American meeting at Cold Spring Harbor it was average; but to a Frenchman it was an insult. I asked Luc eight questions, all scientific," Gallo, who is not noted for his tact, explained.

But they were tough questions because Gallo was being criticized by some of his associates for having helped get the French paper, which was perceived as being rather "soft" (without all the technical questions well explained), published in *Science*. Gallo says he and his people were surprised at the French refusal to collaborate, but essentially "we said oh, to hell with this; we'll do it on our own."

The encounter marked the beginning of the escalating animosity toward Gallo by some in the Pasteur group. This rift extended even to the relationship between the National Institutes of Health and the Centers for Disease Control (CDC—the surveillance agency of the U.S. Department of Health and Human Services), which almost immediately sided with the French. At the time, the CDC was facing a drastic cutback in funding and had to vie with the relatively autonomous National Institutes of Health for the paltry amount appropriated by the federal government for research on the new disease.

Adding insult to injury, the major medical reporter of *The New York Times*, Dr. Lawrence Altman, had been a CDC employee for many years, and that influential newspaper's slant was staunchly French and CDC–oriented. The science press in general followed by interpreting AIDS news with a spin toward one or the other "side" of the issue.

Serious questions of "ownership" of the virus emerged as both the U.S. and French groups developed blood tests for the presence of the virus. After two years of acrimony, these disagreements were settled by an unusual agreement forged at the highest levels of the respective governments. All in all, however, the time and energy consumed by the contretemps would better have been turned toward the pressing needs of research. Unfortunately, such scientific arm wrestling is not unknown when the ultimate payoff will be great. In this case, not only was there a unique disease under way, but it involved a novel category of virus.

Human Immunodeficiency Virus

During the first several years after AIDS appeared, what general public interest there was in the disease revolved around the many titillating aspects of the affected populations. Meanwhile, despite the infighting of various research institutions, extraordinary progress was made in identifying the new virus and understanding many of its peculiarities.

The retroviral family to which human immunodeficiency virus (HIV) belongs, although known in animal diseases, had only a few years earlier been shown by Gallo to infect humans. But the two viruses then identified, human T-cell leukemia virus (HTLV-1 and HTLV-2) were known to cause unrestrained cell growth (as in leukemia), while HIV instead kills cells. HIV was briefly mistaken for its close relatives because some of the envelope proteins are shared, which causes both types of viruses to register positive on tests for HTLV.

Retroviruses received their name because, with RNA rather than DNA as their genetic material, they must go through an additional step to reproduce themselves. (Viruses such as polio, measles, and mumps also have RNA as their genetic material but do not go through the processes used by retroviruses.) Retroviruses require the presence of the enzyme reverse transcriptase to facilitate the translation of their RNA to DNA, which then produces a messenger RNA that is the final step in viral replication. Like other retroviruses, HIV causes latent infections and directly destroys relatively few cells. But the AIDS virus preferentially infects cells bearing the CD4 marker, and therein lies its grim ability to cause this unique disease.

When lymphocytes, or white cells, destined to become T cells leave the bone marrow and migrate to the tiny thymus gland, they are programmed to take on a distinct function by virtue of the molecular markers that develop on their surfaces. Two of the types of lymphocytes are programmed with markers designated CD8. This identifies their functions as either killer or suppressor T cells. Killer cells attack invading organisms directly; suppressors turn off the immune system attack once the danger has passed. But the T4 cell carrying the CD4 marker has a regulatory function that is central to the function of the entire system.

Like the conductor of a symphony orchestra, the T4 cell signals when the music should start; since its orchestra members include the antibody-producing lymphocytes, when the T4 cell malfunctions, along with many other immune responses, production of antibodies is also dysregulated. Over time, T4 cells gradually decline in number until the immune system's function is virtually nil. Almost every aspect of immunity is thrown out of balance.

Early in the AIDS epidemic, Allan Goldstein, head of biochemistry at George Washington University in the District of Columbia, noted that AIDS patients were producing abnormally large amounts of the hormone thymosin α_1. This hormone, secreted by the thymus gland, had been discovered twenty-five years earlier by Goldstein and Dr. Abraham White and is an essential part of an effective immune response.

In examining AIDS patients, Goldstein noted that the thymosin was "a peculiar kind." It turned out that a string of thirty amino acids (a peptide) belonging to the hormone and one of HIV's peptides were identical. Goldstein's tests showed what seemed initially to be high levels of thymosin, which would have been protective against the virus. However, after more refined analysis, the scientist found that he was measuring levels of one of HIV's internal proteins. The thymus gland itself has subsequently been found to be eroded in AIDS patients, probably because of molecular mimicry in which the immune system mistakenly attacks the look-alike part of the thymus. The gland is unable to carry out many of its essential functions, and this probably further disrupts the production of competent T4 cells.

Once infection has occurred, HIV is thought to lurk unnoticed for an undetermined period of time, perhaps for as long as twelve years. In the meantime, the virus is passed from cell to cell as infected cells make contact with uninfected cells that have the CD4 marker. This mode of cell-cell transmission prevents the virus from being exposed to antibodies in the blood that might destroy it. Infected cells may still function normally or may be partially disabled, but not until they are activated by some signal—probably to fight off another infection or a combination of infections—does HIV begin killing in earnest.

As the infected, activated cells begin to divide, the virus hiding in them bursts through their envelopes, destroying the cells and spewing new virus into the blood circulation.

Tissues other than lymphocytes also have the CD4 marker; it is present on certain cells of the brain and the colon and studs the surface of macrophages and their first cousins, the Langerhans' cells that underlie the skin. These probably are safe houses for the virus, places where it is protected from antibodies. Hidden in the very cells that should destroy it, HIV plays a cruel game.

But even stem cells in the bone marrow, which are the precursors for many mature cells and as far as is known haven't been programmed with any specific markers, may be filled with the virus.

Because the virus's message has become part of the cells' own genetic code, viral progeny are produced with each cell division. Over time, the burden of infection grows geometrically. At some point it becomes too great for the immune system, and the process that leads to AIDS begins.

The particular illnesses that ensue are determined in part by the presence of other infections. These infections probably play a role in the differences in life expectancy of various people who contract the disease, because it is these infections, not HIV directly, that finally kill them.

Other factors may influence when full-blown AIDS will develop, such as poor nourishment, the isolate (specific makeup) of HIV with which one is infected, age (older people and infants fare less well), and perhaps to some extent the patient's underlying genetics or the route of infection. A transfusion of infected blood or the transplantation of an infected organ, for example, immediately exposes millions of lymphocytes to HIV, while a single virus accidentally injected in a laboratory accident may be eliminated.

Is the HIV virus in itself deadly? Would an otherwise infection-free person living in a sterile bubble ever progress to AIDS? Maybe not, but the death and derangement of critical cells caused by the virus set in motion other processes that may be equally central to the decline of patients. In fact, some physicians feel that autoimmune processes instigated by HIV may be the ultimate insult in a long chain of events (Chapter 15).

By whatever mechanisms, both exotic and ordinary infections have their way in a body with few defenses left. One opportunistic infection after the other makes its appearance as the other infectious agents with which we normally live in equilibrium are freed to multiply.

Many bizarre dysfunctions cloud the AIDS picture. Dementia and extreme wasting are common. Unregulated B cells produce antibodies against many pathogens to which one has ever been exposed, creating immune complexes, the combinations of antigen and antibody found in severe rheumatoid arthritis. Because the immune system serves as a sentinel against cells with the potential to create cancers, once it becomes unable to function, these also appear.

7

Proving the Impossible

The virus-cancer scenario began in 1908 when two Danish researchers showed that serum from chickens with leukemia could cause the
same cancer when it was injected into healthy chickens. Two years
later, Peyton Rous of the Rockfeller Institute in New York was able
to transfer tumors called sarcomas from sick to healthy chickens.
Since these cancers were transmitted by cell-free fluids, the assumption that viruses were involved made sense to the researchers.
However, this claim incurred the disapproval of the rest of the
science community to the point where Rous, at least, backed away
from his cancer-virus theories. Not until 1966, when he was eighty-
five, was Dr. Rous's discovery finally acknowledged with the Nobel
prize.

During the first part of the twentieth century the intellectual
climate of science was heavily weighted against such a discovery.
The little that was known about either cancer or viruses seemed to
militate against the inspired leaps of imagination of Rous and, later,
Ludwik Gross and others. At that time the technology that enables
one to see a virus was lacking, and cancer didn't fit into any reasonable scheme of "infection."

Because some cancers run in families, because others could be
induced in animals by radiation or by hormones and chemicals,

because the term *cancer* describes many very different types of disease, virus-cancer theories just didn't seem to make sense. When the whole scientific community turns its back on an idea, only a very independent worker with independent funding has the emotional and financial wherewithal to pursue it.

There's also the two-way street of animal research. While a popular proposition can be reinforced by the results of animal research, when the same kind of experiment seems to support an unpopular or novel idea, it's easily dismissed as an artifact of laboratory manipulation. Such artifacts are encountered for many reasons; for instance, injecting a germ into an animal approximates but doesn't duplicate a naturally acquired infection, and the transfer of a mature cancer from one animal to another only mimics the natural development of that cancer.

Even today, when a number of viruses have been shown to play a major role in causing cancers, skepticism still exists. Michael Bishop of the University of California at San Francisco has posed the central dilemma. Since no one can explain how a normal cell controls its growth, "it may seem foolhardy to think the abnormal rules governing the growth of a cancer cell can be deciphered." But it is in the very machinery of the normal cell that the potential for cancer exists.

The first oncogene discovered by Bishop and Harold Varmus was originally worked out in the Rous sarcoma virus. Starting with the idea that the virus carries and inserts a cancer-causing gene in chickens, the researchers instead found that the guilty gene belongs to the cell itself: The virus merely picks up the gene from the infected cell. Chemical carcinogens may cause mutations in the normal version of the gene—now called a proto-oncogene—or the virus may insert its passenger gene in the wrong place in the cell's DNA. Also, a switched-on oncogene can disrupt the normal messages of other genes within the cell and thus scramble the directions for cell growth.

Some applications of this recent discovery already have been put to clinical use. For instance, if tissues from some cancers show many copies of the oncogene, physicians know that the cancer has a potentially worse outcome. Aggressive treatment is called for in such cases. Within a few years, probes that recognize turned-on oncogenes should produce early, accurate cancer diagnoses.

By 1960 there were plenty of examples of laboratory animals developing cancers when infected with retroviruses, but they were dismissed as not applying to humans for a variety of reasons. Many lab animals are inbred, and this tends to distill genetic tendencies so that certain strains of mice become supersusceptible to a particular disease, murine mammary cancer, for instance. The famous nude mouse, a wrinkled, hairless, white prune of a creature, has been inbred so that it is born without a thymus gland and therefore has no immune function. Even a grafted patch of black fur is acceptable because without immunity, there are no cells to reject the foreign tissue.

Inbreeding may also pass along to offspring bits of viral information that have been incorporated into the parents' genomes, by ancient infections, for example. In some mice and chickens retroviral infections can become part of the germ cells (egg and sperm), which the offspring inherit without any signs of infection.

However, there had long been hints that more than just lab artifacts might be happening to animals. In the United States research on a cancer-retroviral relationship focused almost entirely on the lab mouse and on viral sequences that were thought to be remnants of ancient infections. It was generally accepted that something was activating normally silent strings of viral information to cause cancer. The molecular biology employed to show this was elegant, but the endogenous (from inside) virus-cancer theory didn't pan out in a literal sense (i.e., inject a virus, grow a cancer), and a generation of American virologists missed the boat. Only a very few scientists were at all interested in the possibility that an animal, particularly a human, might later develop cancer as a consequence of a viral infection.

In Scotland, however, there was someone who was deeply interested. He was a tall, peripatetic, veterinary-trained researcher named William Jarrett, who had worked in East Africa for years and had seen naturally occurring cancers appearing in patterns suggestive of the spread of an infectious agent. In the 1970s Jarrett proved that cats can "catch" and infect one another with a virus that causes leukemia; he named it feline leukemia virus (FeLV).

By 1972 only a couple of American labs believed that humans would ever be found to have retrovirus infections, much less a virus-associated cancer; some claim there was too much emphasis on mouse models. But both Sol Spiegelman at Columbia and Gallo undertook to show that these viruses do cause infections in humans. "It was a rough time because we were very much alone," recalls Gallo.

Building Blocks

What is so routinely assumed by today's researchers about the central roles and functions of T cells and B cells in the immune system was still largely unknown in the 1960s. It was understood that the white blood cells—lymphocytes—are responsible for immunity, but few researchers suspected that these seemingly identical cells have very specific and very different activities. The origins of these differences began to come clear in 1961, when experiments on newborn mice showed that removing the thymus gland can render these animals unable to fight off the effects of viruses and fungi, entities we now know are recognized and dealt with by the T-cell arm of the immune system.

When a part of the bone of chickens, the bursa, was removed, the fowl were unable to make antibodies—a B-cell activity. Apparently the same basic cell was migrating from the bone marrow to be programmed for separate functions by two different systems—the bursa and the thymus gland—though formal proof didn't come until 1970. The details of the interrelatedness of different lymphocytes were still fuzzy, and there was no way to grow them in culture.

After he finished his medical training in 1965, Gallo went to the National Institutes of Health, where his entire professional life has been spent. He reluctantly put in the required time on the clinical wards and in 1966 finally began the research that was his goal. Gallo began by studying lymphocytes, which "wasn't a hot topic then," and learning biochemistry and nucleic acid sequencing. Though he says he was adrift at the beginning of his work on lymphocytes, Gallo steadily grew more interested in how someone develops leukemia and what the molecular basis of this uncontrolled growth of white cells could be. Could it be explained by a single mechanism, or were there multiple causes?

"I didn't think humans were ever going to have cancers caused by viruses. The thought seemed a little bit primitive," he admits.

By 1969 a series of influential ideas had begun to change his thinking. Gallo became friends with Robert Ting, who had worked at the Massachusetts Institute of Technology (MIT) with Salvatoria Luria, a Nobel prize winner in molecular biology, and at Cal Tech with Rinaldo Del Vecco, whose Nobel prize was awarded for his work on animal tumor viruses. Ting convinced Gallo of the possibility that cells can be transformed by viruses. Through discussions with Ting and his own avid reading, particularly the publications of Robert Huebner, Gallo became aware of the field of retrovirology.

"I began thinking that if those viruses are causing leukemia in so many species, even if they don't cause it in man, what we learn may be helpful in explaining the mechanism. So I decided to spend part of my time looking at retroviruses in animal systems," Gallo explains.

He did not go quietly into the field. Belgian molecular biologist Asne Burny, now a leading expert on bovine leukemia virus, says he first heard Gallo's name while working in Dr. Sol Spiegelman's laboratory at Columbia University in the late 1960s.

"Sol came back from a meeting in Paris and asked if I knew a guy named Gallo. 'Imagine,' Spiegelman complained, 'I was asleep at three in the morning when I got a phone call from someone named Gallo telling me how important it is to look for a leukemia virus and reverse transcriptase and we should collaborate!' "

Burny was with Spiegelman during the years when the Columbia researcher's search for a cancer-causing virus "took a wrong turn."

The wrong turn hung on a grotesquely simple difference in approach. Using a bird virus as their model, the Columbia group "did not use detergent to open the virus." As every dishwasher knows, detergent dissolves fat, and fat is a major constituent of many viruses' envelopes. Deciding to wash cells in detergent seems a peculiar basis on which to win or lose, but of such small differences are Nobel laureates often made.

"Howard Temin at the University of Michigan used detergent, and David Baltimore at the Massachusetts Institute of Technology had chosen to focus on murine leukemia virus, where no detergent was necessary," Burny recalls. Both groups were able to make critical discoveries that Spiegelman's group missed, all for the lack of detergent.

During the 1970s Gallo also got to know Myron Essex at Harvard, "who was coming out of George Klein's lab at the Karolinska in Stockholm," where he'd become involved with feline cancer. Essex went on to Harvard and became an important collaborator in Gallo's search for an AIDS virus.

This kind of network of friendships and collaborations is central to science, where boundaries of any kind are a serious hindrance to fitting pieces together. Although people of other cultures approach problems with different sets of assumptions and in different languages, scientific standards are set on a worldwide basis. Increasingly, the best people in given fields are dispersed throughout the world and rely on the results obtained by distant colleagues to move ahead with their own work. In this fashion, the shared resources of major equipment and technology as well as the small but important tricks developed in different labs can be called on without constant reinvention.

At scientific conferences it quickly becomes evident that what goes on in the formal meeting is of far less consequence than what is being exchanged in the halls and lounges. At Gallo's annual by-private-invitation-only lab meeting, informal but binding deals to collaborate are made as needs become apparent in open discussions of new work. It's in these "family" gatherings that the sheer fun of scientific research is allowed to surface as these outwardly serious men and women stretch and let their guard down. But they never give it all away, never tell everything they know, because science has to be funded, and publication of innovative work is needed to attract new money.

It's a complicated and very human world, and the old-boy network, as valuable as it is, often tends to shut out those it finds unacceptable. For them, the road is inordinately long and difficult.

Although Gallo will be remembered for his codiscovery with Luc Montagnier of the virus implicated in AIDS, his most important work may be the identification by his team of what was first called T-cell growth factor, now interleukin-2 (IL-2). This chemical message is central to our rapidly accumulating knowledge about the immune system and its interaction with viruses. Before the discovery of IL-2, cultures of lymphocytes couldn't be kept alive long enough for the necessary experiments with white cells to be carried out. Without live growths of infected lymphocytes, AIDS would be even more mysterious than it is.

When Gallo's National Cancer Institute (NCI) group started looking at white cells with an eye to their possible transformation by viruses, the accepted theory was based on one major wrong assumption and another that was incomplete. The mistaken assumption was that a virus with RNA as its genetic message can't interact with host cell DNA to cause infections and/or cancer. The incomplete proposition, made by Ludwik Gross, was that viruses leave behind some of their amino acid sequences that can be turned on to cause cancers. Howard Temin proposed an intermediary step between infection and cancer called a provirus, but Gallo says that "the details were fuzzy."

Then Temin and Baltimore independently found an enzyme they called reverse transcriptase, which converts RNA into DNA. Since DNA programs RNA and RNA in turn directs the production of proteins, a major constraint against RNA viruses being able to cause cancer was that they were thought not to have the machinery to program DNA, an echo of the Watson-Crick assumption.

The existence of reverse transcriptase (RT) led to an immediate reworking of the old linear paradigm. It became possible to think of an RNA virus infecting a cell, using its RT enzyme to translate its genetic material into DNA, which then integrates with the host cell's DNA. When this happens, the virus's message is incorporated into the cell's genetic code, which produces more virus as it replicates as well as, in some cases, cells with a potential to cause cancer. This discovery pushed the viral-remnant oncogene theory into the distant background by showing that RNA viruses indeed have a mechanism by which they can, at least theoretically, cause cancers.

It was assumed at the time of its discovery that RT is used only by retroviruses. However, it has been shown recently that a range of viruses and other organisms, such as yeast and, in a far more convoluted way, a portion of the hepatitis B virus's double-helix DNA, also have RT activity. But these discoveries came later.

When the first report about reverse transcriptase was published in 1970, Gallo was investigating the possibility that an enzyme can act as a "chemical carpenter," able to assemble DNA molecules in such a way that the product serves as a promoter of cancers. Reverse transcriptase fell loosely into the same family of enzymes and was then thought to occur only in association with retroviruses, so Gallo began looking for the RT molecule in tumor cells. Five years of work produced tantalizing hints that RT activity was present in some cells from leukemia patients, but the scarcity of the virus the experimenters had to work with made definitive results almost impossible.

It isn't easy to get cells to grow in cultures so that they can be watched and manipulated, and even as late as 1976 no way had been found to keep white blood cells alive. As much as anything, this shows how little is really known about biological processes and why animal rather than computer models are essential for most research on living systems. If you knew what information to program into a computer, you'd probably get the wanted answer. But where do the models come from that can pose the questions to ask, the models that determine the what?

If you want to study a type of cell and there's no way to grow it, you look for the natural substance on which growth depends. At least that's what Gallo set out to do. Other labs had found a factor which, when added to human bone marrow cells, would make small colonies of cells grow. Though it was a major discovery, it wasn't useful for virology, and Gallo's group started searching for a growth factor that would keep lymphocytes alive in liquid suspension. They needed a factor that could produce massive quantities of granulocyte cells, the subgroup of scavenging white cells that had attracted Gallo's attention.

The experimenters found that when granulocytes were combined with phytohemagglutinin (PHA), a plant substance that encourages cell division, the white cells grew and divided for a short time. To their surprise, the culture began producing a novel enzyme that stimulated the growth of the white cell colonies.

"Then lymphocytes became an interesting cell for making things grow. Other people were finding that the PHA-stimulated lymphocyte secreted other molecules as well, so it looked like this was a rich source of goodies, and God knows what you'd find," Gallo recalls.

To test their new finding, the NCI group set up a large-scale project with blood samples from many people. The researchers began isolating lymphocytes and stimulating them with PHA, and, according to Gallo, "looked at the supe [supernatant—the fluid left after the removal of all cells] to see what the hell was in there, looking for anything. My goal was still to identify a granulocyte growth factor."

The cells in the stimulated cultures were assumed to be the rare B cells that are infected with Epstein-Barr virus. This virus has the inherent ability to immortalize the B cells it infects, keeping them alive and growing endlessly.

"No one had ever grown T cells," Gallo states. "It was believed not to be possible."

The cells in the culture failed to produce antibody, the primary product of B cells. To double-check their results, an outside lab was asked to type the cells. The results were the same: no antibody. Therefore, what they had to be growing was the other type of lymphocyte, T cells. The group named the new chemical T-cell growth factor, now known as IL-2.

One of the strange things learned along the way about T cells is that when they become activated, as when they are attacking a virus or bacterium, they produce a range of new molecules on their own surfaces. For example, when T cells are stimulated by the plant antigen PHA, they secrete IL-2; at the same time they develop new IL-2 receptors for the protein on their own surfaces. The IL-2 secreted by the cells binds to its own new receptors, causing them to divide. This redundancy enhances the activity of the immune response. It is a remarkable feedback system: recognize, secrete, create, bind, secrete, and finally divide. But how does this complicated sequence of events relate to the development of cancers?

The First Human Cancer Virus

"Between 1970 and 1978 we couldn't prove anything. We were having some good luck, were sharpening our technology, and getting better at what we were doing, as well as gaining a greater feel for virology," Gallo recalls.

At the same time, while Gallo says he didn't know he was on the right track in looking for a retrovirus-caused cancer in humans, "I smelled it a little. But you can't prove anything until you can get an isolate [virus]," Gallo says of searching for a new virus.

In late 1978 leukemic cells from two patients were cultured with IL-2. As the cells divided, evidence of RT activity was noted, and the investigators saw viruslike particles budding from the cancer cells. When the budding particles were examined, they were found to be unlike any known animal viruses and different from the patients' own genetic material. A new paradigm took shape: an unknown virus seemingly existing only in association with a leukemia cell. By 1981 details of the discovery and the likelihood that the virus plays an important part in causing the blood cancer were readied for publication in scientific journals, the important open forum of international science.

For ten years, without knowing exactly what they'd find, Gallo's team had focused on the biology of human blood cells. Since he was untrained in virology, Gallo says it was "learning by the book—like trying to cook; learning by talking to people in the cafeteria, by going out for a drink, and then doing it yourself, but doing it differently." The outcome of this different way is the first virus—human T-cell leukemia virus (HTLV)—ever proved to cause a human cancer.

"It's the greatest life!" Gallo says, describing what it's like to be a scientist. As the prime example of this great life, Gallo points to Ludwik Gross, "the mouse leukemia virus man who is still working in a little lab in the Bronx VA in New York."

"Here's a guy at the peak of his career because of the simple pleasure of knowing that we have human retroviruses now," Gallo says. "He came to Cold Spring Harbor in 1982 for the first international meeting on a human retrovirus—HTLV—and he was like a child, he was so excited."

"Dr. Gross opened the field of mammalian retroviruses with the mouse leukemia virus in 1953–1954. He loves to come to Bethesda and spend a couple of days. He was telling me, 'You have the torch, you know.' He's passed me the torch," Gallo marvels.

Ludwik Gross had been excluded from and disregarded by the scientific establishment for years. As a young surgeon in Krakow, Poland, in the late 1920s, Gross watched cancers spread from his patients' lips to the adjacent lymph nodes in the same way an infection does and wondered if cancer could be caused by a virus. On a trip to Paris a few years later, Gross discussed this possibility with doctors at the Pasteur Institute. Dr. Alexandre Besredka, intrigued by the idea, asked the young man to come work at the institute, where the emphasis was on developing vaccines.

In Paris, Gross's project was to vaccinate healthy mice against cancer. After injecting small amounts of tissue from cancerous mice into healthy ones, Gross transplanted the cancers to the vaccinated animals. His hope was that if the cancer was caused by a virus, the "vaccination" might prevent it from killing the mice. His occasional success was impossible to interpret, Gross recalls, and "did not bring us much closer to the concept of viruses."

During his years in France, Gross was unable to transmit mammary or sarcoma cancers to healthy mice, but he held to his virus-cancer theories and presented them to several influential medical people on a trip to the United States in the fall of 1938. Because he was a Jew, Gross was becoming increasingly afraid of the Nazi steamroller in Europe and on his trip began making applications to work in America.

His applications for positions at Yale, Columbia, the Rockefeller Institute, the Cleveland Clinic, Hanneman Hospital in Philadelphia, and Memorial Hospital in New York were all rejected. "I was turned down everywhere," he says. "Nobody believed at that time in the existence of a cancer virus."

Gross reluctantly returned to Poland and a few months later found himself "driving a friend's abandoned automobile ahead of the invading German Army" into Romania. By selling the car in Paris to generate funds for passage to America, and through a series of interventions by fellow scientists, Gross finally managed to reach the United States in the summer of 1940. His parents and other family members, who had remained in Poland, died under the Nazis.

Gross says that he was "almost obsessed with the idea that tumors and leukemias were caused by a virus" during the months he waited for his U.S. citizenship and a commission in the army. But he had no thoughts about how he could prove his theory. Then, while he was reading about the familial incidence of retinoblastoma, a cancer of the eye we now know is caused by an inheritable defective gene, "it came to me as clear as a ray of sun," he recalls.

"Suddenly, the thought came to my mind that since it was obviously a tumor, the virus, which theoretically would exist, is simply transmitted from one generation to another." The young woman who would become his wife "told me that I came out of the laboratory with red cheeks. I was so excited, and I was saying, 'I have it! I have it!'" What Gross thought he had intuited was that a virus carrying a cancer-causing gene was passed from one generation to the next, with some unknown event activating it. But Gross also knew that his intuition had to be substantiated by experiments.

Gross's commission as a captain in the medical corps came through weeks after his revelation, and he found himself working in a hospital in Tennessee. There was little time and no laboratory in which to conduct his research, so Gross rounded up mice and kept them in coffee cans. He recalls "keeping the colony going in my room and in my car when necessary, waiting for the war to end." Gross had devised theoretical experiments that he thought would prove his hypothesis, and though he says his colleagues made fun of him, they were impressed with his persistence. At the war's end, Gross was assigned to the Veterans Hospital in the Bronx. With the backseat and floor of his car rattling with canned mice, he drove from Tennessee to New York.

Unable to secure research grants to test his theory, Gross cared for patients and fashioned a makeshift laboratory in the oxygen tank storeroom. Even his request for a few mice of a kind that easily contract leukemia was refused by the head of the Cold Spring Harbor Laboratory on Long Island. After finally obtaining eleven leukemia-sensitive mice from Jacob Furth at Cornell, Gross set out to transfer what he thought was surely a cancer-causing virus to other rodents. For four years his work went nowhere. The rumor in the hospital was that, because "I had crazy ideas, I would be shipped out," he says.

Gross was almost ready to give up when he had another flash of insight. This one came while he was attending a lecture describing the paralysis caused by coxsackievirus in mice. The adult mice with which Gross and most others had always worked were unaffected by the virus, but neonates less than forty-eight hours old became paralyzed. Gross understood why his work had failed.

He rushed from the lecture to his storeroom lab, prepared an extract of cells from a leukemia-prone animal, and injected it into day-old mice. Within three weeks all these mice had developed leukemia. He had successfully transmitted leukemia from one mouse to another, proving for the first time that cancer is directly caused by a virus. "So that was it. I immediately received money for cages and assistants to take care of the animals," he says. After eleven years of struggling with improvised experiments, a year later, in 1951, Gross showed that even the cell-free serum could transmit the cancer.

The scientific community, however, remained unimpressed. As Rous had found before him, "Nobody took me seriously. I was severely, sometimes even viciously, criticized," Gross recalls. His work was characterized as "ill-conceived and not reproducible" in many quarters.

Gross nevertheless continued his research and in 1953 identified a second cancer-causing virus of the salivary glands in some mice. Gradually, with an almost evangelical persistence, by going from one researcher to another—often carrying gifts of mice and distributing his extracts—Gross changed the collective mind of the research community. Prestige, awards, and election to the National Academy of Sciences have followed over the intervening years. Dr. Gross, now in his eighties, still a tall, handsome man, remains an emeritus researcher at the Veterans Hospital in the Bronx, though his lab space and assistants have gradually been whittled away to make room for other researchers.

In his 1976 conversation with Ludwik Gross, French researcher Marcel Bessis speculated with Gross: "One wonders if there is perhaps a young man, such as yourself in the 1930s, who is now presenting or publishing work which appears to all of us as 'ill-conceived and not reproducible work.' "

Human T-Cell Leukemia Viruses

Gross's discovery of virus-caused mouse cancers set the stage for a comparable discovery that would directly link viruses with some types of human cancers. But twenty-five years later, Gallo's team was wrestling with the novel virus they had found in leukemia cells. To find a virus is one thing; to assign it a causative role is quite another. As can be imagined from the previous discussion of oncogenes and growth and regulatory factors, the complex mechanisms by which a virus may start the cascade of biological changes that eventuate in a malignancy present a monumentally difficult puzzle.

Gallo's team began to zero in on various characteristics of the newly identified virus that might encourage malignancies, such as the virus's inability to kill cells. This meant that it could become latent, which may be a prerequisite for a cancer-causing virus. But with what kind of leukemia was the virus involved? Mycosis fungoides and Sézary syndrome, two slightly different types of malignancy, were eliminated when HTLV couldn't be found in cells from those patients.

But Dr. Kiyoshe Takatsuki in Kyoto had recently described a T-cell leukemia (adult T-cell leukemia, ATL) that occurred frequently in the southern islands of Japan. The clustering pattern of cases of this virulent cancer, which kills within months after symptoms appear, suggested that an infectious agent might be responsible. In a series of collaborative studies between the Japanese and American teams which included isolating the virus in Japan, the scientists began to close in on a group of cancers apparently caused by HTLV.

Then researchers at the Tokyo National Cancer Institute worked out the genetic sequence of the virus taken from Japanese patients and found that it was almost identical to the sequence of the virus Gallo had named HTLV. Next, Dani Bolognasi at Duke University isolated HTLV in a Japanese-American with a T-cell leukemia. With a virus in hand and an antibody test to establish whether patients with leukemia had been exposed to it, laboratories across the world began to examine the blood of adults with leukemia.

In London's Hammersmith Hospital, Daniel Catovski found the ATL cancer only in Caribbean-born black people. This echoed the experience in the United States, where ATL is a rare cancer and the occasional patient diagnosed with it is almost always black. Everywhere they looked in populations of American and Caribbean blacks, of South Americans and Africans, HTLV infection was widespread, but a relationship between Africans and Japanese, separated by half the world, seemed farfetched.

Next Dr. Isao Miyoshi found an HTLV-like virus in Japanese macaque monkeys and hypothesized that those animals might be its reservoir. It turned out that the human and monkey strains weren't closely related enough to be transmitted between human being and beast. But Miyoshi's observation provided another piece of the puzzle, and in collaboration with a German group, Gallo found that many species of African monkeys (but not monkeys from Latin America) also had antibodies against viruses that cross-reacted with HTLV. Indeed, the African simian viruses were more like the human virus than were those from Japanese monkeys.

What possible connection could there be between infected humans and animals on opposite sides of the world? And why weren't the people of northern Japan infected, except for the Caucasian

Ainu bear worshipers of the most northerly island? Gallo hypothesized that HTLV could have been brought to the western hemisphere with the African slave trade and to Japan in the sixteenth
century by the Portuguese, depictions of whom—with Africans and
monkeys included—are found in old Japanese paintings.

Other questions and probable answers followed quickly. Because
there were considerable differences within confined geographic
areas, with many people in one town infected with HTLV and those
a short distance away seldom carrying the virus, it could be assumed
that HTLV is not spread by casual contact, such as sneezing or
drinking water. Where there were many cases of the leukemia in a
locale, many of the residents were positive for HTLV, and there
tended to be multiple cases within single households.

Further epidemiological investigations showed that the infants
of HTLV-positive mothers were antibody-positive and that an adult
born in one of these areas, even if he or she emigrated in childhood,
had the same chance of contracting this particular type of leukemia
in later life as did those who remained behind. But as one swallow
does not make a summer, this kind of evidence does not make a
direct virus-cancer link. It was, says Gallo, "a starting point for the
next goal, which was to identify the molecular mechanism by which
the virus causes leukemia."

Hit and Hide Animal Models

Asne Burny is a large, exuberant researcher from Belgium,
described by Gallo as "a very imaginative guy who thinks of how
disease occurs, and a good molecular biologist." For four years in
the late 1960s Burny had worked at Columbia with Sol Spiegelman,
in whose lab the "wrong turn" was made while searching for cancer-
causing viruses. "We were looking for 'the monster,' a small piece
of RNA" which would account for virus-caused cancers, Burny says,
a search that turned out to be fruitless.

On his return to Belgium in 1970, Burny ran into a school friend
who had become a veterinarian and told him that he was searching
for an infectious cancer that occurred naturally in animals. To his
surprise, Burny learned that at that very moment an epidemic form
of leukemia was being investigated in Belgian cattle. The veteri-

narian, who was in charge of eradicating the disease, at once suggested that Burny work along with him.

"Indeed, it was a very interesting system," Burny says of the cattle disease, "with the advantage of a naturally occurring virus-induced cancer in all sorts of animals."

The leukemia had been observed first in cows in 1969 and seemed to follow an infectious pattern, although researchers had been unable to find a virus in the blood of the sick animals. When he did identify the virus, Burny made several observations that have been important in exploring the diverse ways in which other retroviruses cause disease. As with HIV and HTLV infections, bovine leukemia virus (BLV) didn't require mass production of the virus to cause disease; the relatively rare virus-infected cell was simply hard to find.

In addition, *viremia*—the state of actively producing virus particles—goes on for less than two weeks after the initial infection, making the timing of the search important. After that, antibodies against BLV are found in persistently infected animals. With the exception of sheep, which uniformly develop cancer, it appears in only about 10 percent of other infected animals. These are chronically infected but lack any apparent disease. Since the infected creatures have antibodies against the virus, it seemed logical that they would either get rid of it or be protected against its effects.

But they don't, and this is how BLV and other lentiviruses (*lenti* means "slow"), including HIV and the HTLVs, are able to become persistent. Instead of killing the virus, the production of antibodies plays a role in establishing latency and causing much of the damage done by the infection. The groundwork laid by Burny's investigation of BLV has helped answer the crucial question of why antibodies against HTLV and HIV in humans don't prevent illness.

More clues to retroviral infections in humans are provided by the visna/Maedi disease in goats and sheep, which was first identified in the 1950s. When infected with the visna virus, macrophage ("big eater") cells, which ordinarily gobble up cells marked by viral infection, become its targets instead.

Johns Hopkins researcher Openda Narayan found that when samples of blood were examined early in visna infection, effective antibodies were present. However, after months of low-grade infection, the original antibodies were useless against the virus. Evidently, says Narayan, the natural selection process was at work. When the researcher grew infected cells in a culture medium from which all antibodies had been removed, absent the threat of being recognized and killed, the visna virus remained the same: Under pressure to escape from antibodies, the virus mutated to evade destruction. Within a few months, the virus managed to change enough to be unrecognizable by the earlier antibodies.

Burny explains that in bovine leukemia, "when a tumor develops, the virus's footprints are always found," and the tumor always originates from the type of cell originally infected. While the virus is "necessary to start the process, to push things along, at a given stage it's no longer necessary and is probably even dispensable." But the virus has done its work already, in effect creating a cellular climate in which cancer, a multistep process, may or may not occur years later. "The virus looks like a fossil; it's there in the cell but doing nothing," says Burny.

Cancer remains an almost entirely mysterious process that can affect virtually any body system. Even when disease is deliberately caused in experimental animals by a substance known to be carcinogenic, although the outcome—a cancer—is known, the how of the process is not.

It's as if scientists searching for the causes of cancers were looking through a knothole, hoping enough action will take place in their field of vision to explain, to clarify. Many strange and marvelous entities have recently come into their range of vision, but what is happening outside of it is still only an educated guess.

Many chemical events in humans take place in a series of cascades, with one domino pushing the next to act, and so on. This is true of the mechanisms by which blood coagulates and of the complement system involved in the immune activity that participates in antiviral attacks. It seems likely that the genesis of cancer also involves a cascade of chemical events, probably ultimately based in part on the genetic predisposition of the individual.

In viruses, both the specific genes and the way they're arranged have much to do with the way they infect cells. Retroviruses that cause leukemia have two distinct activity patterns: If the virus itself contains an oncogene, every cell it infects (remember that viruses are very picky about which cells they can infect) may become a progenitor of cancer. This kind of multicell infection-replication, which is quite rare in human disease, is called *polyclonal*, which means that several cell types are involved. The majority of human cancers, whether viral-induced or not, stem from a single location in one type of cell and are referred to as *monoclonal*.

The lentiviruses share a number of similar genes. For instance, the virus infecting the leukemia cells that Gallo had seen growing in cultures were shown to be driven by the recently discovered *tat*, or transactivating, gene, which so far has been found exclusively in this novel class of retroviruses. When a way to inactivate the *tat* gene is developed, it should be possible to short-circuit the development of cancer in HTLV-infected people. But that's far down the road.

Human immunodeficiency virus (HIV) made its appearance just a few years after HTLV was found and distracted scientific attention from the cancer virus. But as blood samples taken to test for HIV have been examined, a surprising number of people have been found to be coinfected with both cancer- and AIDS-causing viruses. In the United States and Europe, HTLV is found chiefly in black drug addicts. As in Japan, there is great variation in the incidence of infection even from town to town.

In certain sections of Queens, New York, Newark, New Jersey, and New Orleans, Louisiana, for instance, from 20 to 50 percent of black intravenous drug users are positive for HTLV-1 (the digit was added after Gallo's lab discovered an association between hairy-cell leukemia and a virus close to but significantly different from the first HTLV). There also appears to be a considerable difference between the strains of HTLV-1 infecting Japanese and Americans, with most of the American virus producing a weak response to the test to which the Japanese strain reacts strongly.

HTLV-1 is thought to cause leukemia in only 1 percent of infected people, but since its latency can last as long as thirty years, it's hard to judge the long-range future of infection with this virus. This begs the question: What about prevention?

"This really is an infectious virus disease," states Gallo. "Therefore, prevent it. Find out who the carriers are and get rid of the mode of transmission. That is, don't allow that carrier to put his blood in the blood bank. [In 1988 the Red Cross began screening blood donations for HTLV-1.] Then you think of a vaccine, for the Caribbean, for southern Japan, for all the endemic areas."

But the problem is that one becomes infected early in life, and once the affected cells are transformed, there's probably no need for the virus to spread and replicate. Thus you're back to trying to treat the cancer itself. It becomes equivalent to treating a nonviral malignancy, such as cancer of the pancreas or the breast. "It doesn't do you any good knowing the cause once it's occurred in this case," Gallo complains.

"With HTLV-1 our hope is that one of the genes that's maintained in the parent and daughter cells is necessary for the maintenance of the cancer and we'll be able to intervene with that . . . with understanding. It's possible, however, that the virus is only needed in initiation and doesn't have to go from cell A to cell B to do its job. One cell becoming a neoplastic cell produces descendants that are all cancer cells," he explains.

Currently there is little we can do but look for new viruses, ascertain if they are connected to diseases, and attempt to plumb their mysterious life cycles. Clinical applications will have to wait for this, and Gallo feels that effective treatments growing from the understanding of molecular mechanisms are a decade or two away. The probable scenario is that there will be steady progress over a period of time, and, he says, "*then* there's going to be an adaptation in therapy that's going to be phenomenal." With a knowledge of what part viruses play, of their interface with the growth and regulatory factors of cells, therapy will be able to change its present scorched earth mode—chemotherapy, surgery, and radiation—for another that specifically targets only cells with the abnormal growth patterns of cancer. The sketch for such therapies is already on the drawing board.

Retroviruses as Therapy

Since changes of one kind or another in the genes of cells that become cancerous are responsible for their unregulated cell growth, identifying these changes and correcting the defective genes would be the ideal therapeutic approach. Consequently, the rapidly expanding field of human gene therapy is being viewed both as a potential method for correcting the more than 2,000 known inherited disorders and as a new avenue to cancer therapy. What better vehicle to install corrected genetic information than a virus whose ability to target specific cell types and reprogram cellular genes is already well established? HIV provides a clue to how this might happen.

HIV is especially well endowed in the gene department. As well as having all the genes found in other retroviruses, HIV has at least four that are unique: *Sor, 3′ orf, art,* and *tat.* The *tat* gene was discovered by William Haseltine of Harvard, who showed that when *tat* is removed from HIV, the virus ceases to reproduce. Theoretically, by finding a way to block the *tat* gene, one could prevent the replication of HIV. Alternatively, if one created a mutant HIV from which *tat* had been removed, with the virus's other attributes such as being able to target and enter cells left intact, the mutant would be able to target cells and perhaps turn off the growth signal. Another AIDS-related problem, Kaposi's sarcoma, that seems dependent on an unregulated growth signal enhances this idea.

Kaposi's sarcoma (KS), the most frequently encountered neoplasm in HIV-infected patients, is not a standard kind of cancer. It does proliferate in an uncontrolled way, but KS is created by growth of the very small blood vessels. These vessels are made up of vascular smooth muscle cells that normally grow at an extremely slow rate and seldom proliferate. Except during pregnancy, when new blood supplies have to be created quickly, the turnover rate of the smooth muscle cells is "100 to 1,000 times slower than bone marrow cells," according to Judah Folkman of Harvard.*

But in HIV infection, an abnormal growth factor is released by the KS cells themselves. This overwhelms the functions of an unidentified factor that regulates normal growth of blood vessels. It must be important, because in Gallo's lab nude mice injected with the supernatant from KS patients develop KS of their own cells. Since there is an animal model to work with, it should be possible to find and restore the missing factor.

A comparable loss of cell growth regulation is caused by many congenital genetic defects, such as the absence of a single gene that creates severe combined immune deficiency syndrome. Without this functioning program to direct the manufacture of the enzyme adenosine deaminase, children are born with a condition much like AIDS.

By coupling a normal gene with a retrovirus—minus the virus's

---

*Judah Folkman, at the Laboratory of Tumor Cell Biology Meeting, Bethesda, Maryland, August 1989.

gene that causes "infection"—researchers hope to take advantage of viruses' ability to target specific cells and send a normally functioning gene into cells to correct the defect. Even now scientists are using insect viruses called baculoviruses that have been manipulated to carry part of HIV's envelope in the hope that a vaccine against AIDS can be made in this fashion.

Within the lifetime of most readers a new paradigm for the detection and prevention of many cancers and congenital disorders will take shape, courtesy of the virus. Before long, viruses will be active agents of cure as well as random agents of destruction.

8

The Herpes Family

Who's on First? No, Who's on Second?

Though scientists are able to identify, snip, and sort individual viral genes, though they can map their different regions and generate three-dimensional pictures of many of their molecular structures with computer imaging, they've failed to come up with a simple way to classify viruses.

The effects of viruses were known long before such an entity was identified. Jenner and Pasteur were vaccinating against smallpox and rabies without having seen the agents that caused those diseases. As these agents came to be recognized as entities causing different illnesses, viruses were initially named for the diseases with which they were related, for instance, mumps, measles, polio, influenza, and hepatitis.

Other viruses have been named for the locales in which the diseases they cause first surfaced, such as Lassa fever virus from Nigeria. Some, such as herpesviruses, were assigned to a category determined by superficial appearance or other points of similarity, though even they often are given names associated with their discoverers or activities. An occasional virus is simply identified by the site it infects, its most obvious effect, or both; for instance, RSV (respiratory syncytial virus), attacks the lungs and causes cells to form syncytia, clumps of cells in the shape of crescents.

New imaging techniques have made it possible to assign viruses to categories that reflect their genetic material, DNA or RNA, single- or double-stranded. But even this doesn't serve well as a way to categorize them, because many of these newly designated pathogens have habits which require special notice, such as being "slow" or oncogenic (cancerous) or causing latent or persistent infections. The activities of a few viruses seem to include virtually everything that a virus is known to do.

But if a researcher's name and accomplishments are to be in the history books, the naming of the pathogen, vaccine, technique, or disease he or she discovered or first wrote about is the best insurance. The names of Salk and Sabin, of Creutzfeldt and Jakob, of Marek, of Epstein and Barr will be as recognizable to future generations of scientists as Proust and Hemingway are to readers of fiction.

While we don't want to become embroiled in the problems of scientists who are constantly trying to decide under which often arbitrary classification to list a virus, the importance of naming a discovery was never clearer than in the controversy surrounding the naming of the virus that causes AIDS.

HTLV-III/LAV/ARV = HIV

When Luc Montagnier's team in Paris first saw what they thought might be a virus implicated in the emerging disease, they called it LAV (lymphadenopathy-associated virus) because it was taken from the swollen lymph node of a man in the early stages of the illness. At the same time, Montagnier suggested that it belongs to the newly discovered family of human T-cell leukemia viruses, but none of this was certain until Gallo's group characterized (looked at the genetic makeup of) similar viruses from other AIDS patients. Because he had discovered the HTLV viruses, Gallo rechristened his first HTLVs as I and II and changed leukemia to lymphotrophic, referring to the white cells thought to be the new virus's target. He called the AIDS-associated virus HTLV-III.

Not everybody was happy with that. It got so a scientist or science writer could declare which side of the ensuing controversy he or she was on by referring to the virus either as LAV or HTLV-III or by combining the two. Then Jay Levy in San Francisco found what turned out to be another isolate and named it ARV, for AIDS-related virus.

Many scientists, trying to stay neutral, wrote about this infinitesimally small bit of stuff in papers that were hard to read because of the numerous times one had to climb over the acronym HTLV-III/LAV/ARV or, in the case of most French and some other scientists, LAV/HTLV-III/ARV.

After three years of acronym wrestling, the international scientific community got irritated and convened a nomenclature conference from which issued a simple solution: Call the thing HIV for human immunodeficiency virus. In relief, which one assumes was felt most acutely by the typesetters of medical publications, HIV was quickly adopted. But that was just the beginning.

Even before the identification of the human immunodeficiency virus, researchers in two experimental primate centers had discovered that some captive rhesus monkeys were dying from an illness with almost identical symptoms. The virus they subsequently identified was named SAID, with the S standing for "simian." SRV-I (simian retrovirus), which, though apparently genetically unrelated to SAID, caused the same kind of disease in macaques.

Other monkey viruses followed, and their acronyms had to be augmented by tiny subscripted letters such as *agm* and *mac* (for African green monkey and macaque) as similar viruses began to surface in other types of primates. And the west coast of Africa produced more of the human retroviruses. HIV became HIV-I, II, III, or IV, but no one was quite sure which referred to which, and Montagnier insisted on calling the one his team found in Africa LAV-2.

Talks at scientific conferences concerning viruses are largely extended slide shows. Presenters rushed to conform their nomenclatures to the new ones, but it was impossible to keep up. In late 1987 in New York, a major AIDS researcher sought to relate one of the new West African immunodeficiency viruses to the others that were already known, explaining how it was closer to the monkey viruses than it was to the original HIV. Above the diagonal strips used to show genetic comparisons of the numerous viruses, past and present initials were crammed together with slash marks setting them apart from one another.

In the darkened conference room the audience strained and

squinted, but it was impossible to read and remember what the multiple designations referred back to in the time each slide could remain in view. When the presenter used the phrase "originally called" for at least the tenth time, the audience, out of nervousness or confusion or because they were privy to the political implications of these acronyms, began to titter.

Personalities and nomenclature aside, where does HIV best fit into these categories? It's certainly capable of causing acute infection, as demonstrated by the flulike episodes often experienced when a person is first infected. But it's also slow, and shares a considerable molecular relationship with other slow viruses (lentiviruses) such as the visna virus that infects sheep. HIV is persistent (once infected, always infected) and even has some characteristics associated with viruses described as oncogenic (cancer-causing).

Because AIDS is so new and deadly, one might assume that it's unique in the world of virus, and in some ways—particularly in regard to its apparent lethality and its numerous genes—that's true. But there are older viral infections of humans which, although they don't kill as spectacularly, are as enigmatic as HTLV-III/LAV/ARV/ HIV, etc.

All this goes to say that if the reader has a preconceived notion of what herpesvirus is, it's best to set it aside before going on, because the common understanding associated with herpes won't be much use. A few years back, fever blisters came to mind when we heard herpes; then genital herpes infection became an important consideration among the sexually active. But that's only one branch of a family of viruses that have far more serious effects on human health.

Somewhere Between an Art and a Science

When French epidemiologist Guy de Thé showed President Georges Pompidou an electron micrograph of the Epstein-Barr virus, "he looked at me amazed and said, 'I would never believe something so beautiful could be the cause of cancer.'" De Thé replied, "Ah, they give cancer not for pleasure; it is just a by-product."

Guy de Thé first became interested in herpesviruses through his interest in Africa and his relationship with Dr. Denis Burkitt.

"Burkitt is not a scientist; he has no degree, he was just a surgeon, and a military surgeon, which for a Frenchman means he's low grade," explains de Thé about the (to Americans) peculiar hierarchy of European degree status. De Thé said this not as a criticism but as an enormous compliment to the doctor who attempted to understand and treat the sometimes unique diseases that affect Africans.

"Denis has all the qualities of a good observer: He saw a subtle thing and tried to understand it, to pose the right question," explains de Thé. For instance, wondering why in Africa there was no cancer of the colon, "You know what he did?" de Thé asks.

"Only Denis Burkitt could do it. He weighed the feces! At the end of twentieth century he weighed the feces!"

Burkitt looked at the volume of feces Africans excreted after eating to find out the time the feces took to traverse the alimentary system and concluded that a slow transit time plus a low volume of feces played a major part in causing cancer of the colon. "They said he was completely mad! That this was like science in the Middle Ages. But, of course, he was right!" de Thé chortles.

He was right enough that his conclusions led to the fiber-rich diets, which produce rapid excretion, promoted as protective against cancer twenty years later.

Guy de Thé has traveled with Burkitt in Africa and across the world in his role as a viral epidemiologist. He describes time spent in the field, sleeping in muddy tents and eating half-cooked goat and suspect grain mush, with glee.

"What I like about epidemiology is that you have to be a detective. You have to be all the time like a radar, trying to find, to understand," he explains.

Eighteenth-century Africa was considered a privileged place in which a young researcher could study to his or her heart's content infectious and parasitic diseases but not cancer. The continent was thought to be virtually free from the neoplasms that affected the western world. However, it gradually became clear that this apparent difference was due to the burden of disease, particularly diarrhea-causing infections and malaria, which, with malnutrition, caused death before Africans reached the age when cancers ordinarily appear in other populations.

As various European nations established colonies and, along with their abuses, brought medical and public health interventions which lowered the infant mortality rates and lengthened the average life span, and as scientists developed techniques by which even unusual cancers could be identified, these killer diseases were added to the list of Africa's health problems.

In 1946, fresh from World War II service as a surgeon, young Denis Burkitt let his deeply held religious convictions lead him where he felt he could do the most good. Garden of Eden though it is in many ways, East Africa exacted an immediate toll: The Burkitts' baby son died of dysentery on the ship out from England.

The burden of sickness was appalling. Stationed in Uganda, Burkitt found an infant mortality rate between 250 and 300 per 1,000 births and great morbidity and death caused by liver cancers following hepatitis B infection. Patients with kwashiorkor, the sign of advanced protein starvation, flooded the hospitals. Babies routinely died from diarrhea, measles, and polio. Burkitt had anticipated all this, but he soon saw something totally new to his experience: the dreadful cancer of the face and jaw later to be named after him, Burkitt's lymphoma.

As chief of service in the department of surgery at Mulango Hospital, serving under Sir Albert Cook, who was then head of the Ugandan health service, Burkitt was called into consultation to see a five-year-old boy named Africa who had a disfiguring cancer of the jaw. Burkitt operated on the child but was unable to save him. Soon afterward, at a hospital on Lake Victoria, he saw another child with the same enormous tumor, and he wondered.

Searching the archives of the Mulango Hospital, Burkitt found numerous references to this monstrous growth and presented them at a regular Saturday morning meeting of hospital physicians. The radiologist said he had recorded seventeen such cases. Burkitt began collecting tissues and sending them along to colleagues in Europe, hoping that clues to the children's cancers could be found. But before light would be shed on the tumors Burkitt was seeing in Uganda, a circuitous research route covering three continents would have to be traveled.

The Creeping

For several hundred years, eruptions that creep in a line along the skin were attributed to filterable agents, that is, to viruses. As different as their effects on health may be, morphologically the herpes (Greek for "creeping") family of viruses have certain characteristics, such as double-stranded DNA and a protein coat, and replicate in similar ways. Otherwise, they're not that closely related except in one important way: Herpesviruses can all become latent. How a virus becomes latent rather than causing acute infection, along with how and why a latent infection is activated, is a major mystery that is a tremendous stumbling block for the development of effective antiviral therapy.

Shingles, a painful nerve disorder that usually affects the skin of older people, is caused by infection early in life with herpes zoster, the cause of chickenpox. But the virus "goes latent in the nerve ganglion and sits there during generation after generation" of cell division, explains virologist Martin Hirsch of Massachusetts General, who has specialized in this family of agents.

"Immunosuppression induced by chemicals or that which comes with AIDS, or trauma to tissue, or sunlight or wind or emotional stress can all activate the virus, and it comes back down again," Hirsch says. This suggests not only that we don't know what activates a latent virus but also how little is understood about what "stress" is.

After being hidden and going unnoticed for as long as sixty years, as those with genital herpes well know, the virus can become activated on a regular schedule.

Five or six human herpesviruses among the sixty or so that infect animals have been identified so far, and many think this is just the beginning. The herpes family of viruses is widespread throughout the animal kingdom, infecting the highest and the lowest. Herpes causes a range of illness from encephalitis of the newborn, to diseases mostly mediated by the immune system like mononucleosis, to cancers that are now known to be herpes-induced. Many of these cancers occur in animals, such as monkeys and frogs, but twenty-five years ago there was no proof of a similar association in humans.

The Henles and the Herpes

By the 1950s some progress had finally been made in understanding and treating cancer. That, plus insights into the probable role of immune system cells in mediating health, was compelling enough for a search for causes of human cancers to be undertaken in earnest.

"In earnest" in contemporary research largely means money.

Where 100 years ago Louis Pasteur, a chemist by training, could, according to Hilary Koprowski, "dabble in sancrosanct preventive medicine" and come up with a vaccine against rabies, today even an inspired conceptual leap has to be backed by an infrastructure too great for any but giant corporations or governments to fund.

In fact, the very act of securing funding requires great expenditures of time and money. This slows research and creates frustration and bad-blooded competition between researchers who might otherwise be collaborators. Though the system works in many respects, in other ways it tarnishes what could be a shining example of human cooperation.

It was no wonder that in the late 1950s, when Gertrude and the late Werner Henle of the Children's Hospital of Philadelphia received a telephone call from the assistant director of the National Cancer Institute inviting a grant application, which was to be merely a brief outline—because, as Henle wrote, "detailed proposals were not required because there were no leads on which to base them" —they responded eagerly. The best part of the offer was that the funding included ten years of committed support.

"Can you imagine," Henle rhetorically queried researchers at a 1984 symposium, "ten years without writing a competing grant application?" The Henles' stream of research had to do with the interplay between the immune system and viruses, and so they "were not unprepared for the telephone call" from Washington.

Most grants have to be renewed after a couple of years, and because researchers generally go after the unknown, it's often hard to justify the millions of dollars required. It certainly wouldn't do to send in a proposal that stated, "We have this hunch," although it would often be the most accurate reason for heading in a particular direction. In the 1950s a relationship between virus and cancer in humans was altogether speculative, although Ludwik Gross had shown that viruses cause mouse cancers.

Since arriving in America in the 1930s, the doctors Henle had spent their entire professional lives at the Children's Hospital, where they performed much of the basic research that led to vaccines for mumps and influenza. Henle described his and his wife's subsequent discoveries as having "depended on being at the right place at the right moment, on recognizing and grasping an opportunity as it arose," and on "just good old plain luck."

One element of this luck was the longtime presence at Children's Hospital of the surgeon in chief, C. Everett Koop, who later would be named surgeon general of the United States. From Koop, Henle had secured numerous cells from childhood cancers against which to test his lifelong theory that a chemical produced by a (maybe) cancer-causing virus could cause interference between itself and normal cells, short-circuiting the virus's otherwise destructive potential.

To test his interference theory, Henle had been working with vesticular stomatitis virus (VSV), a deadly virus of hooved animals. According to the researcher, "We had still to find a single cell culture of man or beast, fish or fowl, reptile or insect that was normally resistant to this omnipotent virus." The children's cells supplied by Koop were no exception. The fact that they were cancerous didn't prevent them from being attacked and destroyed by VSV.

Then Koop returned from a conference in Africa and shared with the Henles his suspicion that the spread of Burkitt's lymphoma in children in tropical Africa seemed to follow the pattern of a transmissible agent.

"We immediately wrote to Denis Burkitt and other physicians on the African scene," Henle recalls, only to learn that many others were already deep into various projects to explore this cancer.

It appeared that all the possible bases had been covered already by investigators seeking to identify what was causing these neoplasms in African children.

"We apparently had missed the boat," Henle recalled.

But the sailing was merely delayed, perhaps to take on the passengers that held essential parts of the puzzle which would rapidly come together in the Henles' hands. In the doctor's own words, a fortuitous sequence of events then occurred.

A year after the Henles' disappointment, Dr. Anthony Epstein and his colleagues at the Middlesex Hospital in London established continuous cultures of Burkitt's lymphoma cells. On electron microscopic examination, they found herpeslike virus particles in a small proportion of the cultured cells. This discovery aroused few ripples of excitement because most virologists assumed that the virus was herpes simplex, cytomegalovirus, or chickenpox virus, and none of these or any animal herpesvirus known at that time was thought to cause cancers. The virus indigenous to Burkitt's lymphoma cells was therefore thought to be a harmless passenger of no particular interest.

"As a last resort, Tony Epstein sent the Burkitt cell cultures for identification of the virus to Klaus Hummeler, who recently had spent a sabbatical in his laboratory and had been in charge of our virus-diagnostic service at the Children's Hospital," Henle explains. But that lab was being dismantled because the Pennsylvania Health Department had withdrawn its support for financial reasons.

So, without a facility in which to even look at the cultures, "Klaus Hummeler came to our office, waving the bottles, to ask what should be done with them."

Gertrude Henle responded, " 'Give them to me,' and thereby came our chance to work, after all, on Burkitt's lymphoma," Henle explains.

With the rescued cultures, Gertrude Henle was able for the first time to demonstrate several important things. The cancer cells from African patients, unlike any other cells with which they'd worked, were highly resistant to the lethal vesicular stomatitis virus. This appeared to advance the Henles' theory that some interference ("or interferon") was being produced by the Burkitt's lymphoma cells.

To follow up on their suspicion that there might be a novel virus in the Burkitt's lymphoma cells, the Henles needed fresh blood from an African with the illness. Some things seem predestined.

"We learned that a Nigerian Burkitt patient had been flown to the Clinical Center at the National Institutes of Health in Bethesda for plasmapheresis [blood washing], and attempts were being made to sediment [by centrifuge] the expected C-type Burkitt's lymphoma virus from the plasma by high-speed centrifusion. It is truly amazing," said Henle, "how many 'viruslike particles' can be sedimented from anybody's plasma, but C-type virus particles have not been among them."

Henle at once telephoned the director of the National Cancer Institute, Ray Bryan, to request the supernatant—the leftover fluids from the patient's blood.

"My God," responded Bryan, "they pour them down the drain!"

But one more plasmapheresis run was planned, and the valuable material was collected and sent to Philadelphia. When the supernatant was tested against cells, the researchers showed that it acted just as the samples from England had. The new virus was lurking in the fluid.

"The resulting euphoria was of only short duration, however," recalls Henle, because the same tests of the blood of healthy Americans showed that it also contained the pathogen.

Though they were absolutely certain that what they had was a new virus, one that was implicated in Burkitt's lymphoma, how could it be proved when the same pathogen was evidently ubiquitous in the blood of Americans who never develop the cancer?

Could the new virus be related to one of the known herpesviruses? Henle's lab tested it against herpes simplex, cytomegalovirus, varicella, and various animal herpesviruses, but no relationships between them were found. Provisionally, they named the new virus EB virus, after the sample in which it had first been found, which had been sent to them by Drs. Epstein and Barr.

"Helper viruses had just become fashionable," said Dr. Henle, "but those we tested, that is, herpes simplex, mumps, and reo-3 viruses," had no effect on the new agent.

At the same time they were trying to establish a causal relationship between the Epstein-Barr virus and Burkitt's lymphoma, the Henles were finding antibodies against it in blood samples from everywhere: the Amazon jungle, South Pacific atolls and Aleutian islands, and down the road from their laboratory. Crowding and poverty seemed to be related to a high infection rate with EBV, as did infection with the virus causing polio, though this was not necessarily related to contracting the disease itself.

Among the samples against which the new virus was tested, those from Africans with nasopharyngeal carcinoma surprisingly acted just as did the sera of Burkitt's lymphoma patients. Thus there seemed to be yet another cancer with which EB was associated. But if EB caused such cancers and almost all adults everywhere had been exposed to it, what was the common illness it had to be causing in the American population? This time the Henles "were assisted beyond the call of duty by one of our young female technicians."

In research labs, when blood or tissue samples are needed, they're usually taken first from whoever is available, and lab workers are the most convenient donors. So when the young graduate student became ill, her blood status was known, and the EB virus, which hadn't been present in her earlier blood samples, was found.

"This was not a laboratory infection," Henle insisted. "Our technician was a very pretty girl and was thus exposed to the kissing disease virus by the natural route."

The Kissing Disease

As innocent as it seems in today's sexual climate, where words formerly never spoken aloud are casually bandied about in TV ads, when the "kissing disease" was diagnosed back in the 1940s, sick college students were embarrassed to own up to having mononucleosis. Though its cause was unknown, mono represented a potentially serious though usually temporary sickness accompanied by sore throat, fever, and other symptoms. However, it was best known for the utter fatigue that accompanied it. Mono primarily affected people in their late teens and early twenties, and when it started in a closed environment such as a college campus, it raced from one dorm to another.

But it takes more than a single case of sickness in a graduate student to document the cause-and-effect relationship needed for scientific proof. From researchers at Yale, the Children's Hospital team obtained blood samples taken from healthy freshmen as they entered school, with new samples being drawn and examined each of the following years they spent in college.

Without knowledge of which blood samples were from students who developed mono during their college careers, the Henles were readily able to identify those who had become infected. The same thing happened with the students' blood that had happened with that of the lab technician. When a sample "converted" from having no EBV to being infected with the virus, mono followed; without the virus, there was no mono.

The field opened quickly once it became possible to test for the new virus. Traces of EBV were found in 90 percent of adults worldwide, but whereas only 26 percent of Yale freshmen were positive, in Africa 95 percent of children under two years old had already been infected. Yet there seemed to be virtually no mono among Africans of any age, and the very young western children who were infected experienced a usually insignificant set of symptoms or nothing. Like polio, hepatitis B, and many other viruses, whether one became ill depended largely on one's age at infection.

EBV infects B cells and reproduces mainly in the epithelial (skin) cells of the salivary glands and probably in the oropharynx, which lies back of the nose. The uniqueness of EBV rests on its ability to cause some infected white blood cells—which otherwise survive only a generation or two—to reproduce indefinitely in the laboratory. Like cancer cells, cells infected with EBV become immortal.

But what correlation could there be between the lethal cancer of African children and mononucleosis, the kissing sickness of largely upper-class western adolescents? De Thé subsequently found that the 20,000 annual cancers that grow behind the nasal cavity in Chinese are strongly linked with EBV. Are genetic traits involved? Is it the age at which the initial infection occurs?

Environmental cofactors such as malaria in African children have been implicated in the development of Burkitt's lymphoma. Spices used to pickle vegetables in some areas of China have been tentatively suggested as playing a role there. However, proof of these possible cofactors is still tenuous.

Another possible outcome of EBV infection was thought to be the cancer called Hodgkin's lymphoma. In nearly all cases of this disease, the malignant cells have markers associated with the activation of lymphocytes, indicating chronic infection. In addition, EBV genetic material is often found within the cancerous cells. Although this does not constitute proof that EBV is directly implicated, it at least suggests a possibly important role of EBV.

While most adults have been exposed to EBV and carry small numbers of immortalized B cells, it requires a possibly enormous number of variables—from age at infection to cofactors such as food intake, genes, genetic markers, and immune suppression—for an ordinary infection to cause such a range of illness. However, the new herpesvirus HHV-6 has also come under suspicion since it was discovered a few years back.

De Thé says, "The fact is that we have many, many carcinogens in our environment, and we know that you need a stepwise progression between normal and cancer cells. For each step you have a fight or play between the causative agent and the susceptibility of the individual. So when you realize that it's maybe many different genes controlling each disease, you realize that the chance to get a cancer is really low. And any little change in your lifestyle may be enough to completely change your risk." If only we knew what those changes really were!

EBV is only one of the herpesviruses whose range of effects on humans are far from being fully known. But that's the case with many viruses whose subtle influences may have dramatic consequences down the road. No better example of this exists than what happened following the great influenza pandemic that swept the globe just as World War I was drawing to an end.

Influenza: A New Paradigm

In 1985, researchers completely mapped and created a three-dimensional model of one of the viruses that cause the common cold. This was the first time an animal virus had been visualized in such detail; scientists called it a remarkable feat of research that was performed using equally remarkable techniques. The virus, rhinovirus 14, a member of the picornavirus family, was first crystallized. Then, using an atom smasher, an intense beam of x-rays was directed through the crystal. The beam, deflected by the virus's structures, produced hundreds of images of every atom making up the virus.

The photographs were broken down into 6 million bits of information that were fed into a Cyber supercomputer, which in about a month produced a facsimile of the virus's complete structure. The soccer ball–like rhinovirus has twenty triangles fitted together; between the joins are deep canyons whose joins apparently are too close together for antibodies to get in and prevent infection from occurring.

As potentially vital as such models will be in determining viral structure, understanding the relationship of structure to the way each virus infects body cells, and learning how immune system molecules neutralize viruses, the average person is still going to have cold symptoms six times a year. High tech notwithstanding, the 200 to 300 distinct viruses that produce cold symptoms will probably always outsmart our best efforts by sheer weight of numbers. The influenza virus, on the other hand, usually manages to evade our most advanced techniques because of its ability to constantly recombine its genes.

An Epidemiological Rosetta Stone

The discipline of epidemiology began as a method for keeping track of outbreaks of infectious diseases. Today its role has been expanded so that epidemiological techniques are applicable to a far broader range of health problems. The activities of medical detectives are now used more often to identify patterns of illness and to compare the rates of occurrence of different diseases and causes of death in various populations.

In large medical institutions epidemiological techniques are essential to identify the frequent outbreaks of nosocomial (hospital-caused) infections. On fortunately rare occasions, epidemiologists are called in to establish unusual patterns of death in order to catch a nurse or orderly who has been playing God with patients' lives. Vaccination programs are often predicated on epidemiological accounting methods which show whether prevention makes sense in terms of cost: How much illness will be prevented, how many lives will be saved, and how many people can be expected to be injured by a mass vaccination campaign?

More esoteric studies are carried out by creative epidemiologists. For example, in 1980 a young trainee at the Centers for Disease Control in Atlanta found that there is no advantage to trying to escape a tornado by driving in a direction perpendicular to the fast-moving wind tunnel, folklore to the contrary.

Any unusual distribution of illness or one that doesn't seem to follow a known course attracts the attention of disease detectives. The outbreaks they track are often small. Several years ago many teenagers in a rural midwestern town were reported to be suffering from severe gastrointestinal upset. They hadn't shared contaminated food at a church picnic, and in fact most of the sufferers weren't even acquainted with one another. If it wasn't food, what could have been causing the continuing problem? Contaminated marijuana: The grower was storing his stash in a chicken coop where small quantities of the fowls' infected feces had mingled with the weed.

"Quick and dirty" surveys are always hoped for by epidemiologists. In these cases, they can rush in, identify what's causing the illness, and recommend a remedy. Legionnaires' disease is a perfect example. It was initially identified in a distinct group—American Legion conventioneers staying at a single hotel in Philadelphia—and the epidemiologists could strike while the iron was hot. A formerly unidentified bacterium was found growing in the hotel's water storage unit and spreading to rooms through air-conditioning units. The bacteria was eliminated, and that was the end of the miniepidemic. It also alerted the medical community to other outbreaks possibly caused by the newly found pathogen, which has subsequently been found to contaminate water in other large holding tanks.

With Legionnaires' disease there was little to confound the relationship between cause and effect, no long passage of time during which other bacterial or viral infections or environmental or personal events confused the picture with extraneous but potentially important "noise." But when additional elements are added, as is usually the case, the clues tend to lead in and out of dead ends while the epidemiological picture becomes more complex.

Dr. David Fraser of Swathmore College wrote about a severe flulike illness that affected employees in a hospital in Rochester, New York, in the early 1970s.* It didn't take long to identify rodent lymphocytic choriomeningitis virus in the blood of several of the sick workers. A review of hospital records revealed that of the twenty-three cases, eighteen were people who worked in the radiation department. When the radiation area was established as the epicenter of infection, it was simple enough to examine the hamsters caged there. It turned out that as part of cancer experiments, the animals had been implanted with tumors and that choriomeningitis virus had infected the cancerous tissue. A large percentage of the healthy workers who had handled the animals were antibody-pos-

*David Fraser, "Epidemiology as a Liberal Art," *New England Journal of Medicine* (September 16, 1989).

itive, meaning they'd been exposed to the virus. But samples of blood drawn from a number of employees with no connection to the radiology department also were found to be antibody-positive. How had the photocopier repairman and a plumber become infected?

The same way as the lab workers: by breathing air contaminated with droplets of hamster urine. Because of space constraints, the radiology department's photocopying machine had been stuck at the end of a room with access through a narrow passage flanked by the animal cages. The eventual conclusion of the epidemiologists was that infected cancer cells had been implanted in healthy hamsters, which had contracted the viral illness and, as is the case with many viruses, had begun shedding live virus in their urine. Droplets of urine containing the virus were being inhaled by all those who passed between the hamster cages. Nine months later, a national outbreak of the same infection in hamster owners was traced to the part-time breeder who had supplied the Rochester hospital with the tumor cells.

Although not generally thought of as a high-tech area of science, epidemiology has gradually adopted more sophisticated technologies. These technologies make it possible to look at subtle influences on health, such as the effects of environmental poisons on body systems, or to track a batch of contaminated milk or meat from an infected animal through its processing and ingestion by a consumer. This is accomplished by following the special genetic signature of the pathogen collected from infected animals in the field and tracing it through butchering, distribution to stores or restaurants, and into the bodies of consumers.

Without question, the fast work of Centers for Disease Control (CDC) epidemiologists during the first months of the development of AIDS hastened its identification as an illness caused by a transmissible agent rather than a novel environmental contaminant or a disease such as Legionnaires' contracted through casual contact. Long before the actual human immunodeficiency virus (HIV) was seen, its presence was suspected: The clustering of cases early in the epidemic had the look of person-to-person contagion by a virus. Despite a lack of knowledge about what actually was causing the disease, screening requirements for blood donation were changed.

Recently, we've discovered that humans are infected with other slow viruses. These infections greatly extend the period of time through which epidemiologists have to search for causes. Instead of asking sick persons where they've been recently or what they had for dinner last night, information has to be developed on the basis of recollection and long-lost or forgotten medical histories before an association between cause and a possible eventual effect can be made. This isn't easy. It's hard enough to recall what you ate for dinner last week, let alone a transitory bout of illness experienced as a child.

Epidemiology, says R.A. "Rai" Ravenholt, "operates most effectively when one is measuring relationships between distinctive disease phenomena and causes closely related in time and space." Nevertheless, Dr. Ravenholt undertook to decode the effects of the great pandemic of influenza that occurred in the early 1900s. He set himself this task because its effects, he suspected, were still being felt fifty years later and also because it was simply there, a challenge to Ravenholt's medical and intellectual curiosity.

Influenza

Why research an old flu epidemic? Flu is something we "catch" every couple of years. There isn't much to do about it—effective antiviral medicines remain to be developed—so we rest, drink lots of fluids, and take something to relieve the fevers, aches, and pains. Then we go back to work or school none the worse for a week's malaise. The very old, the very young, and those with chronic respiratory illnesses or diseased hearts are urged to get vaccinated to prevent or moderate the illness. There's no possibility of effectively vaccinating massive numbers of people against every flu that comes along, any more than there is of vaccinating against a cold.

Influenza is caused by a myxovirus, of which the different types of the A strain are responsible for epidemics. They aren't as numerous as the assorted cold-causing viruses, but influenza viruses, like the AIDS-related viruses, are marvels of mutation. When you're exposed to one strain, the immune system routinely creates memory cells to protect you against the next exposure to the same virus. But since each flu virus is in genetic transit, the virus's genetic code reads that particular way for only a short period of time. Within a few

years the virus shifts its program so that the immune system doesn't
recognize it; another bout of sickness follows exposure to a virus
with a slightly but importantly different set of markers. Between
pandemics, which occur about every ten to fourteen years, various
interim strains of flu virus cause relatively moderate illness.

Researchers in New York and Wisconsin recently studied the
changes of one gene in the A strain of fifteen influenza viruses that
had been documented over a period of fifty-three years. Tracking
what is called the NS gene, the scientists found that its evolutions
provide "a good molecular clock" that can be used to anticipate the
appearance of a new strain.

The observation that pandemics occur every ten to fourteen
years is the kind of scientific "fact" that seems relatively simple on
its surface. But beneath this statement of fact is a question that
beggars the imagination: What unrecognized patterns underlie a
predictable recombination of genetic material? Researchers call it a
good molecular clock. Do molecules, then, also have regulatory systems? Is there an inherent order at this level of the universe?

The fact that the virus shifts its recognizable elements means
that vaccines have to be changed every couple of years to keep pace
with what's coming around the corner. Like a soup pot into which
a bit of this and that is thrown, as a new viral variant appears,
scientists toss a killed version of it into a basic vaccine mixture,
hoping the immune system will find something to recognize and
attack. This type of vaccine is called *multivalent*. However, the major
concern is that a virulent strain, such as swine or Hong Kong, will
put in another appearance.

The influenza virus's uniquely segmented RNA, which is different from the neat, intact strands of most other viruses' genetic
material, probably lays the groundwork for its remarkable plasticity.
This peculiar configuration, which schematic drawings often depict
as looking rather like a pasta salad, is thought to enable the virus
to shift its genetic makeup to escape destruction by antibodies. But
this is not just a minor shift of a couple of molecules on its surface
or envelope like the disguise techniques used by most other viruses:
To successfully evade detection, the influenza virus becomes another
virus. This pathogen must have an enormous reserve of genetic
material, because so much seems to be expendable.

Another indication of the lavish habits of this myxovirus is that when a new brand of flu spreads through a population, the older variety simply disappears. For instance, when Hong Kong flu killed 80,000 people in 1968, the virus that had caused Asian flu in 1957 no longer existed. It had been overtaken and eliminated by the new model.

The genetic changes of influenza viruses account for differences in their virulence. Although virtually every winter arrives with some flu bug in tow, most of these viruses don't cause widespread or serious illness. But every so often something seems to click into place in the viral genome that causes it to become transmitted easily and to have dangerous effects on many of those who become infected.

In experiments with a variant of reovirus type 3, which ordinarily infects the neurons of the brain to cause fatal encephalitis, the alteration of a single gene prevented the virus from reaching the brain. Reovirus has been manipulated in another way so that the amount of virus available (through shedding of virus from tissues) to pass the infection to another host is reduced.

Another insight into shifting virulence with genetic changes comes from a recent experiment that combined two nonvirulent herpes simplex viruses. The recombination generated a virus that was lethal to the brain cells of mice. Fortunately, not all viruses play this bait and switch game: Rabies vaccine is still based on the viral strain discovered by Pasteur in the late 1800s, and the yellow fever and polio vaccines retain their effectiveness despite being based on viruses isolated fifty years ago.

Reservoirs of Virus

The viral diseases of humankind that we have the least chance of eradicating are those which are maintained in other animals or are very easily passed by casual contact. For example, because only humans are susceptible to measles, that disease could be eradicated just as smallpox was by a global vaccination program to drastically diminish the number of people who can transmit the infection. With a small enough pool of potential hosts, a disease will disappear.

In diseases where another animal is involved in the chain of transmission, there's little hope of eradication. The malaria parasite is such a disease. Wild monkeys are infected with malaria, and mosquitoes carry the parasite back and forth between primates and people. We can kill the mosquitoes and their larvae by using pesticides around the home, but there's no way to eliminate the disease in monkey habitats. Some day a vaccine may be developed to protect individual humans, but the malaria vaccine will be needed as long as mosquitoes bite infected monkeys. The eradication of influenza is even less likely.

The constant genesis of novel influenza viruses is probably caused by the pathogen's ability to infect an unusually broad range of domesticated animals, some of which can transmit the virus directly to humans. Numerous strains of human flu viruses have been found in animals, including horses, turkeys, ducks, and pigs. An ancient Oriental farming practice might long ago have supplied a natural test tube environment in which the flu virus learned to accommodate to a range of hosts by using pigs as a mixing vessel.

Traditionally, new strains are first identified—sometimes as long as a year before they reach our shores—in countries rimming the Pacific basin. In those countries, and particularly in southern China, polyculture is an age-old practice. In this type of farming (called the "blue" revolution), the ponds in which fish and ducks are kept are fertilized with pig manure. The pigs are susceptible to and become infected with both the duck and human flu viruses and pass them between themselves and humans.

The different types of flu that circulate across the world each year are usually named after some place in the Orient: Taiwan in 1986, Shanghai in 1987 and 1988, and so on. Looking ahead to the wave of flu expected in 1990, the groups responsible for devising a vaccine decided to include the current Taiwan strain of virus, to substitute Shanghai for Sichuan, and to replace the present B strain with a new type of B isolated the year before in Japan. This potpourri of viral strains may or may not serve to outwit whatever new combination is to be served up by the Orient.

As the population of the developing world expands, enormous pressures for increased food production are encouraging cocultural agricultural practices. In Thailand, a kind of apartment-house arrangement is used in which hens' cages are hung above those of pigs, which feed on hen droppings; the pigs in turn are caged above ponds, which are fertilized by their feces.

Many African countries are also beginning to experiment with fish-bird-pig polyculture. The health implications of this economical way of fertilizing with fresh pig manure and thus increasing the production of badly needed protein are potentially serious if at the same time it encourages the development of even more novel flu strains. A case in point is the virus implicated in the 1968 pandemic of Hong Kong flu, which had a gene that scientists discovered had come from an avian influenza virus normally found only in waterfowl.

Although these mixtures of flu-carrying animals are probably the staging ground for major changes in the virus, even at home we may pick up flu from our animals. In 1982 turkeys were found to be infected with a virus indistinguishable from the human and pig viruses when a technician handling the virus developed flu. The virus isolated from her throat was identical to that which was infecting the turkeys. In 1988 a pregnant woman died of flu after she had attended a state fair and visited the pig barn, where the animals were suffering from flu.

The range of influenza is constantly being found to be wider than was thought. Blood tests of beached and dying pilot whales off Cape Cod in 1986 showed that the animals were infected with flu; in 1980, 500 Atlantic harbor seals died of a virus that is closely related to influenza. The 16,000 or so seals that died in European waters in 1989 were found to be infected with a virus much like that which causes distemper in dogs.

It is the potential reappearance of swine flu that most worries epidemiologists. Predicting that the strain that caused the 1918 pandemic was due to resurface in 1976, the federal government sponsored the development of a vaccine and mounted a massive immunization program. But the flu never materialized, and some people contracted Guillain-Barré syndrome (a progressive weakness caused by changes in the central nervous system that sometimes follow recovery from infectious diseases) from the vaccine.

President Gerald Ford and his health officials were bitterly criticized for the expensive and in some cases dangerous vaccination program, but they could be forgiven their concern: In the years after World War I, the greatest epidemic ever known had swept across the world, and it seemed to health officers that a repeat of that disaster was in the making. The following examination of the influenza pandemic will show why medical decision makers were pressed to come to the conclusion that if the approaching epidemic was caused by the same virus that appeared in 1918, the risk was worth taking.

"Sleeping" Sickness

In 1918 a novel influenza virus collided with humankind in a killing spree. Half the people of the world were affected, and 22 million died. Many of those who seemed to have recovered from the several waves of flu succumbed to encephalitis lethargica, named by Constantine von Economo and commonly called sleeping sickness. We think of AIDS as a unique situation because it apparently sprang full-blown from nowhere, but the same was true of this strange brain disease. It came, killed and disabled millions, and then disappeared. But while this widespread horror hasn't come again, outbreaks of similar localized encephalitis lethargica had been noted for hundreds of years.

Sicknesses with the same symptoms and outcomes were described in 1726 in London and in 1891 in northern Italy, but the twentieth-century epidemic was first recorded in 1915 in Romania. The illness raced across Germany, crossed to France, and spread through Europe, reaching London in 1918, where it was called epidemic stupor. When the influenza pandemic began in 1918, encephalitis lethargica, which by then had run its course on the Continent, returned in an even more virulent second wave.

Despite the confusions of World War I, Dr. von Economo meticulously described the infectious spread and effects of encephalitis lethargica but finally, in resignation, wrote that "we have reached the limit of our bacteriological and biological knowledge" and that a solution would have to "await a fundamentally new discovery." Almost sixty years later, despite scientific advances that would have seemed miraculous to von Economo, there is still no definite proof that the agent responsible for encephalitic brain damage was either a novel unidentified agent or that the influenza virus has the capacity to damage the tissues of the brain. Even less is known about how the virus performs its terrible magic.

Between 1918 and 1920 over half a million people were killed outright by this severe neurological disorder. The survivors created a pitiable reservoir of postencephalitic parkinsonian patients because in many of them the virus sought out and destroyed tissues of the central nervous system and brain. The new syndrome, though similar to the Parkinson's disease seen sporadically in elderly people, produced the same symptoms but with cruel exaggerations.

As brilliantly described by Dr. Oliver Sacks in his book *Awakenings*, it seemed as if a "thousand new diseases had suddenly broken loose." Writing of the bizarre effects of encephalitis lethargica on individuals, Sacks runs a litany of behaviors: "simultaneous and virtually instantaneous onset of Parkinsonism, catatonia ('frozen'), tics, obsessions, hallucinations, 'block,' increased suggestibility or negativism, and thirty or forty other problems."

Some of the sick were thrown into states in which they were, according to Sacks, "unable to work or to see to their needs, difficult to look after, helpless, hopeless, so bound up in their illness they could neither react nor relate . . . these patients were put away in chronic hospitals, nursing homes, lunatic asylums, or special colonies, and there, for the most part, they were totally forgotten—the lepers of the present century: there they died in the hundreds of thousands." Sacks goes on to say that the myriad manifestations of this single disease were like "a Hydra with a thousand heads."

A physician describing Danish schoolchildren who survived encephalitis lethargica wrote in 1924 that "changes in morals and character are the features which primarily stamp the picture, frequently revealing a curious uniformity in different patients. At school, owing to their lack of concentration, their unreliable faculty of perception and retention of ideas, their failing interest in their work, and perhaps primarily on account of their lack of mental perseverance, these children became backward."

Although von Economo had concentrated on the Parkinsonlike outcome of the epidemic, psychiatrist Karl Menninger had reported that "one hundred cases of mental disease associated with influenza in the recent pandemic have been studied at the Boston Psychopathic Hospital. Eighty of them have been extensively analyzed. The range of mental disturbances is wide. For convenience they are readily classifiable into four groups: delirium, dementia praecox, other psychoses, and unclassified. Of these, the second [dementia praecox, now called schizophrenia] is the largest group numerically."

Fa-aniniva: A Strange Legacy

Although he had read of von Economo's disease in medical school, Rai Ravenholt gave it little thought during his years with the CDC as a distinguished international epidemiologist. Then, in 1982, while attending a conference on measles, Ravenholt had reason to delve into memory and bring what he knew of encephalitis lethargica into the present. To him there was a logical connection to be made between rubeola, or "red" measles, and flu: Both belong to the myxovirus family of viruses, and in very young children measles sometimes causes subacute sclerosing panencephalitis (SSP), a severe neurological disease that degrades the brain and usually leads to death.

Ravenholt realized that the long period of latency between measles and the development of SSP might also hold true for the epidemics of encephalitis lethargica (EL) that apparently followed those of swine flu. This was an association that Ravenholt says had too long been "relegated to an intellectual ash heap."

"I thought I should be able to go back and look at the death records systematically and lay them out to see if there was a characteristic lag between the pneumonic peak and the occurrence of any encephalitis lethargica," he explains.

Reasoning that if EL is indeed related to the influenza pandemic, there should be an echo, Ravenholt determined that every time there was the burst of pneumonia, a wave of EL would follow. To test this, he began looking at death records.

"There is a wonderful history contained in old death records. A mute eloquence of earlier days of storm-swept Puget Sound, of falling Douglas fir, of avalanche and cave-in, the rush for gold in Alaska, accompanying miseries of diphtheria and encephalitis, of syphilis and murder," Ravenholt rhapsodizes. "You look at death records and see all this. It mirrors the evolution of society."

A few days spent in the archives of Seattle examining the causes of the death and illness that followed the 1918 flu epidemic convinced him he was on the right track. Ravenholt had recently read about the swine influenza epidemic in a book by Dr. Alfred Crosby titled *Epidemic and Peace* and said that the thing that caught his attention was Crosby's mention that the only place in the world that didn't have flu was American Samoa, "though Western Samoa had it in spades," Ravenholt recalls.

It was a ready-made laboratory situation: two isolated island nations only fifty miles apart and identically peopled. Even the instant of infection was known; it happened when the ship *Talune*, with flu aboard, put into port in Western Samoa on November 7. The consequences were grave: Within two months the island nation had lost 8,000 people, one-fifth of its population, to influenza. Because word of the galloping epidemic on his sister island had quickly been transmitted to the governor of American Samoa, a strict, effective quarantine had been imposed there.

In Seattle, Ravenholt immersed himself in the musty death records of 1918–1925. They showed clusters of EL cases occurring about one year after each of the first three peaks of influenza-pneumonia mortality. Traveling south to the U.S. archives at the University of California, he discovered records from American Samoa which "confirmed the widely held belief that epidemic influenza was not evident in American Samoa during the years 1918–1925."

While studying these records, Ravenholt says, he was amused by the "distinctive vicissitudes of Samoan life revealed by deaths from 'a falling coconut,' from 'eating shark's liver,' and from 'gangrene and septicemia from tattoos of thighs and hips.' "

"As near as I could discern from reading back, looking at the attack rates in populations where records were kept, there must have been a million and a half cases of lethargica around the world. About a third of those probably died in the first month or so, and about 1 million recovered. But it is known that 80 percent of them [the recovered] subsequently developed Parkinson's disease, so even today we may have a few cases of parkinsonism in people who survived."

So far, everything he had read buttressed Ravenholt's suspicions that a single viral agent was responsible for what had been thought of as two distinct and different diseases. However, the death records could not distinguish whether the degenerative brain disease was caused by another virus, was due to inheritable causes, or was a consequence of influenza infection. More evidence was necessary, and so Ravenholt left for the Pacific.

His first stop in American Samoa was at the Lyndon B. Johnson Tropical Medical Center in Pago Pago, where he stayed at the Rainmaker Hotel, made famous by Somerset Maugham, "where the rains do come with violent force." Although Ravenholt knew that flu had bypassed the island, he wondered if EL had also. Indeed it had:

The only excess in death for the period was from manava, a gas-trointestinal disease.

Making the short trip across to Western Samoa, Ravenholt delved into its mortality records with the help of interpreters. Writ-ten in Samoan, the early records stated the most prominent cause of death: *fiva* (fever), *tale* (cough), and *faamai-leaga* (fatal disease). He also came across *fa-aniniva*, which means "fatal disease of the head."

The peaks and valleys of each wave of *fa-aniniva* followed by about a year those of illness and death from the episodes of influ-enza, just as he had predicted they would. "As far as I'm concerned, that was just as powerful evidence—coupled with the patterns I got in Seattle—as looking through a microscope," comments Ravenholt.

Many researchers disagree with Ravenholt's conclusion that EL is related to influenza. Both diseases occurred in pandemic form around the time of World War I, and neither has been seen since. However, the discovery of new viruses in recent years and their association with both novel and old diseases have made us begin to consider these pathogens in a broader way. But Ravenholt's prop-osition that the 1918 influenza pandemic planted the seeds for "mental" disorders which crowded hospitals and lunatic asylums for decades extends the possible effects of viral infection far beyond anything we've previously thought.

Even ancient epidemics of influenza continue to intrigue re-searchers. The great plague of ancient Athens described in careful detail by Thucydides raged for three years between 430 and 427 B.C. That disease fits no known constellation of present-day symp-toms, and according to Dr. Alexander Langmuir, the fall of Athens to Sparta was caused by a combination of infectious agents.

Dr. Langmuir, for many years chief of epidemiology at the CDC, wrote of the epidemic, "It fits all the criteria, epidemiologically and clinically, for influenza complicated by toxic shock syndrome." Thu-cydides had written of symptoms that included coughing, fever, diarrhea and vomiting, thirst, blistering skin, amnesia, and gan-grene. Langmuir says that it is the last symptom that provides the clue: Gangrene is a common result not of flu but of toxic shock syndrome.

You may remember a little over ten years ago when the discovery of toxic shock syndrome created a stir and prompted many legal actions by women who had used high-absorbency tampons. As it turned out, the familiar staphylococcus bacterium, when allowed to grow into large colonies—an environment provided by the tampons—secretes a poisonous toxin.

To Dr. Langmuir's way of thinking, the downfall of Athens was caused by the ubiquitous bacteria infecting already debilitated flu sufferers. If these two agents joined forces to create a novel epidemic, is it possible that the 1918 epidemic and the wave of EL that followed it were also caused by double agents rather than by the influenza virus alone?

We may never know whether this is true, but there is no doubt that a virulent strain of the virus can do far more than cause the symptoms of "just flu."

10

A Matter of the Heart

Dr. Ravenholt's interest in influenza is more than the professional curiosity of an epidemiologist. The deaths of two young men during the Asian flu epidemic in 1957 laid the groundwork for another possible effect on health which he is currently investigating. "I think the swine influenza virus, and perhaps others, had a much broader spectrum of action than has been appreciated," he explains. One of these concerns the muscle that is the heart.

As director of the National Centers for Disease Control's communicable disease section in Seattle, Ravenholt was intimately involved when the Asian flu epidemic hit the west coast in the winter of 1957. "We knew it was coming: The Centers for Disease Control and the World Health Organization were tracking it from early spring," he says.

The first hint that another epidemic was in the making came when details of its impact in Hong Kong were reported in *The New York Times* in April. Reading the paper, Dr. Maurice Hilleman, chief of research on respiratory diseases at Walter Reed Army Medical Center just outside Washington, D.C., realized that with 10 percent of the island's population reported ill, the new strain was a serious one.

In less than a month Hilleman had secured samples of the virus, which were quickly turned over to six vaccine makers. By August these labs had produced 15,000 gallons of vaccine. The vaccine, however, was only about 45 percent effective, and during one week in October almost 12 million Americans were sick. Fortunately nowhere near as virulent as the 1918 epidemic, the Asian flu killed approximately 12,000 in this country.

That September, the Seattle medical examiner called Ravenholt to show him pathology slides taken from an eighteen-year-old who had died of influenza. The apparently healthy young man had developed a flulike illness and while walking across the room had simply dropped dead. "I looked at the slides: It was as if his heart muscle had disintegrated," recalls the epidemiologist.

At almost the same time a young cousin of Ravenholt's in Germany on a musical Fulbright developed influenza. When questioned later, his landlady said he had just told her he was over his illness, when he suddenly dropped dead. The postmortem examination showed the same degeneration of heart muscle as had occurred in the young man from Seattle.

Analyzing the Seattle death rates for the last four months of 1957, Ravenholt found that while the number of deaths attributed to flu was relatively small, the rise in total mortality was very substantial. "I remember that we had an excess of 265 deaths in Seattle as compared to the last four months of the preceding year. Most were older people, and most were attributed to cardiovascular diseases, not respiratory," he reports.

"Everyone is aware when the virus attacks the respiratory system: They're coughing, bringing up phlegm, but most are ignorant that similar inflammatory processes are going on throughout the cardiovascular system, that the heart and blood vessels are inflamed and suffering. But they can't spit and they can't cough. Since we can't really monitor the cardiac system the way we can the respiratory, it comes as a surprise when they suddenly die."

Though most deaths formally assigned to influenza are identified as caused by the pneumonia that often accompanies the infection, every time there is a major flu epidemic, cardiovascular mortality escalates. "It is really the thing that causes substantial elevation in cardiovascular mortality," insists Ravenholt. The weakness and lethargy people experience for days to months after a serious bout of flu is a commonsense indication of the probable effect of the influenza virus on the cardiovascular system.

The graph depicting mortality in the United States from 1900 to 1980 looks like a downhill ski slope over rocky terrain, except for the startling upward spike that defines 1918–1920. But when mortality from ischemic heart disease is superimposed on the same graph, it presents quite another picture.

When one looks into past records, deaths from cardiovascular illness are seen to have been sharply declining until 1920, when they began a precipitous rise. In all age groups, starting in 1920, mortality soared and remained extremely high until a decline began in the late 1950s: It is now one-third lower than it was thirty years ago. The correlation between these dramatic differences has led Ravenholt to "consider the possibility that the epidemic contributed to the pandemic of cardiovascular disease in the 1920s to the 1950s, which is waning now."

"That's not genetic; that's environmental," says Ravenholt, who disagrees with those who cite diet as playing a major role in most cardiovascular diseases.

Since the early 1970s death from stroke, which is caused by hemorrhages of small blood vessels in the brain, has fallen 42 percent. Many physicians suggest that better treatment of high blood pressure, changes in diet to lower cholesterol levels, a decline in smoking, and an increase in exercise are responsible for the declines in caused mortality. Though these factors probably have had some effect on stroke and other vascular conditions, the shift is too great to be assigned to gradual and spotty changes in personal behavior, drugs to lower blood pressure and cholesterol levels, or even better care in hospital coronary units.

Dr. Michael DeBakey of Baylor University in Texas, one of the world's leading heart surgeons, was quoted recently (in *The New York Times*, November 14, 1989) as saying that his observations of more than 15,000 patients have led him to conclude that cholesterol is not the central cause of atherosclerotic heart disease. The underlying causes, says DeBakey, are unknown, and although cholesterol appears to play a role, most researchers agree that many other factors are involved.

Indeed, the abrupt and continuing rise of heart disease immediately after the influenza epidemic, coupled with the inexplicable decline beginning in the 1950s—a period of thirty years during which those affected by the 1918 flu would have decreased significantly in number—indicates that there may be a relationship between the two, according to Ravenholt. Certainly no one could claim that stress has declined since the beginning of the Cold War or that diets were significantly changed during those years. In fact, the national public health education program against cholesterol began only in 1985. Many of you will recall that the "ideal" recommended diet until recently included dairy and animal products —milk, eggs, butter, cheese—which only recently have fallen into disrepute.

Looking only at the recent trend toward a lower rate of heart disease without taking into account the literal epidemic of coronary heart disease that began abruptly around the time of World War I is a bit like arguing over the cause of a broken leg while ignoring the fall that preceded it. Whether that special virus played a role may never be determined, but to Ravenholt the drastic change in the patterns that followed the epidemic are compelling: The crucial value in documenting such late effects of influenzas may be crucial because it is a preventable disease.

In fact, there *wasn't* much coronary heart disease around before 1920. In a lecture in 1912, the celebrated physician Sir William Osler (who offered the ultimate prescription for the common cold: "Treat it," he said, "with contempt") told his audience that a doctor in training would go years before seeing a patient with angina, which is pain caused by a deficient blood supply to the heart. Dr. William Heberden (after whom the Heberden's nodes of osteoarthritis are named), who died in 1801, also considered this symptom of heart disease a rare condition.

Ravenholt speculates, "One may be talking about a long-latent interval, about people who may have been severely injured in childhood or even in utero, and who may die forty or fifty years later" from heart-related diseases. Surely if a flu virus can damage the heart muscles of young, healthy people, causing them to simply drop dead, its effects on the cardiovascular system of those who survive the initial onslaught may well be long term and subtle. Subtle they may be, but both clinical experience and experimental work show that viruses can and do cause serious changes in the health of the heart.

Viruses and the Heart

Catherine Fabricant of the New York State College of Veterinary Medicine writes, "Our studies have established that a herpesvirus infection leads to atherosclerosis strikingly similar to the human arterial disease."* The researcher concludes that this virus may play a primary role in a disease previously considered to be degenerative or metabolic in origin. Fabricant has been able to show that when normal chickens are infected with Marek's disease herpesvirus (MDV), they develop lesions in blood vessels identical to those seen in humans.

To test her theoretical proposition, Fabricant divided a group of chickens in half, with one half receiving a cholesterol-poor diet and the other half receiving a cholesterol-supplemented diet. Then half of each group of the fowl were infected with MDV. Regardless of diet, none of the uninfected birds developed arterial disease.

In contrast, both groups of birds infected with MDV developed lesions in their arteries. According to Fabricant, these lesions "closely resembled human atherosclerosis in character and distribution." The lesions or plaques were primarily composed of the cholesterol and saturated esters that accumulate in arteries to cause human heart disease.

Putting theory into practice, Fabricant's group administered the vaccine against MDV to chickens and repeated the experiments. Even with high cholesterol intake, none of the vaccinated birds developed accumulations of cholesterol plaque.

Following up on Fabricant's results, cardiology researchers from Houston's Baylor University compared 157 men undergoing vascular surgery for atherosclerosis with a control group (comparable ages, backgrounds, etc.) whose members had high cholesterol levels but no signs of heart disease. The tissues removed during surgery were examined for the presence of three herpesviruses, HSV 1 and 2 and cytomegalovirus (CMV).

The results showed no difference between cases and controls vis-à-vis the first two viruses, but significantly higher levels of CMV antibodies were found in the vessels of the surgical patients. Commenting on the result, the researchers proposed that "since CMV

*Catherine Fabricant, "Herpesvirus-Induced Atherosclerosis," in *Concepts in Viral Pathogenesis* (New York: Springer-Verlag, 1984).

antigen, but not replicating virus, is found in cells cultured from the walls of atherosclerotic vessels, the artery itself may be the site of viral latency."*

It may be, then, that replication of CMV within the muscle fibers of blood vessels creates sites in which cholesterol accumulates to form plaque. Persons who have undergone bypass surgery to replace clogged vessels with "clean" ones are likely to reduce their subsequent cholesterol intake nearly to zero. Nevertheless, within ten years fatty plaques tend to redevelop in their arteries.

CMV infection is almost universal in adults, but because the virus remains in a latent state, symptoms of active infection are rare. Exceptions are seen in immunocompromised patients, such as people with AIDS and those receiving transplanted organs. But CMV appears to have more tricks than those of other reactivated herpesviruses.

For one, CMV shares certain similarities with the class I human leukocyte antigens that in large part are responsible for recognizing viruses. The HLA genes, which are present on all cells, are also the primary target when rejection of transplanted tissues takes place. In studies of chronic graft-versus-host rejection of bone marrow grafts, CMV infection was found to precede the condition, often by months. In addition, the higher the rate of infection with CMV, the greater the numbers of HLA receptors to attract the cells that cause rejection.

The exact role CMV plays in this process is uncertain, but a new infection with CMV or reactivation of the virus in patients who have had organ transplants definitely affects the rate of success ("taking") of a transplant. Researchers have reported that patients who have never been infected with CMV often receive transplanted livers contaminated with the virus. The infection is then "almost invariably associated with clinical features" such as hepatitis, pneumonia, and eye infection, say the scientists.

In transplant recipients who are already infected with CMV, the latent virus may be reactivated or reinfection with the viral strain carried by the donor may occur. Reactivated infection doesn't produce symptoms, while infection with CMV from the donor organ

*J. Melnick et al., "High Levels of Cytomegalovirus Antibody in Patients Requiring Surgery," *The Laucet* (August 8, 1987).

does. In other words, the antibodies against CMV that are already in place are able to control the virus, but the same antibodies have little effect on the new strain of virus introduced with the organ transplant. Several studies have emphasized a link between rejection and CMV.

While the association of CMV with the development of heart disease and transplant rejection seems farfetched, it hints at the multiple possible roles of the same virus in diseases with extremely different outcomes. The fact that a certain pathogen usually produces certain symptoms means that, given other circumstances— such as the underlying genetics of the person infected or cooperation with another agent—an unusual outcome may follow. The virus that is responsible for an illness may also be different enough from the usual strain that a disease of a different nature can ensue.

Virus Variants and Clinical Disease

The ability of many viruses to mutate during an infection may change the course of an illness. In some instances the virus becomes less virulent to infected cells. Consequently, the cells display no signs of infection and the immune system ignores the presence of the virus, allowing it to become persistent. Another possible outcome of virus mutability is that antibodies made during a previous exposure may not recognize and defend against another virus from the same family because of a slight difference in strains.

The paramyxoviruses include the viruses that are responsible for mumps and measles. Strains 1 and 2 of another family member, the parainfluenza virus, cause one in three cases of childhood croup, which in adults manifests as bronchitis. Experiments with strain 3 of the parainfluenza virus have shown that there are three separate variants. One results in acute illness with atrophy (shrinking) of the lymphoid organs (spleen, tonsils, thymus, etc.) and death. The two other variants produce a chronic condition that causes hydrocephalus (accumulation of fluid around the brain).

The reoviruses that cause mild respiratory infections also have three different variants. When injected into mice, a variant of type 1 infects neurons and causes lethal encephalitis, while another variant causes hydrocephalus; the third is a mild-mannered variant.

The coxsackievirus family (named for Coxsackie, New York, where the virus was first isolated) consists of two distinct strains. The A strain is most frequently associated with mild upper respiratory illness, but the B strain—which infects cells of the gastrointestinal tract—is associated with several acute diseases, such as meningitis, rashes, and pericarditis and myocarditis, which are inflammations of the tissues surrounding the heart.

In a single individual infected with the coxsackie B4 strain, researchers at the National Institutes of Health found different markers on the same virus when it was isolated from different organs. In other words, once it was in the body, the virus "spun off" variants that were able to infect totally different organs. As we have seen before, the presence or absence of specific surface receptors can influence both which hosts and which cells within a given host can be infected by a particular virus. Coxsackieviruses and influenza viruses can bind to a wide variety of cell types from different species. That ability and the mutations of the virus within the host make the outcome of infections with these agents problematical.

Some members of the coxsackievirus family can produce clinical effects ranging from muscle weakness to paralysis. In infants, strain B viruses (which usually infect cells of the stomach) may result in feeding difficulties, fevers, and other symptoms, including various signs of cardiac distress. These problems can lead to death from collapse of the circulatory system. In older children and adults, the virus also causes cardiac symptoms, and strict bed rest is sometimes necessary to control tachycardia (erratic, rapid heartbeat) or heart failure.

In mice, depending on the variant, coxsackie B virus may cause myocarditis or encephalomyocarditis (affecting both brain and heart), or it may cause nothing.

Ravenholt calls the 1918 influenza epidemic "an epidemiological Rosetta Stone" by which several subsequent upward surges of disease can be analyzed through a consideration of the possible role of the flu virus as a contributing factor. Not only coronary heart disease patterns seem to have been influenced by the pandemic. For instance, cancer of the stomach, the incidence of which has decreased greatly during the past fifty years, showed a sharp increase in the ten years after 1918. Was this also influenced by influenza?

H.L. Mencken expressed incredulity at the lack of interest in the influenza pandemic when he wrote, "The epidemic is seldom mentioned, and most Americans have apparently forgotten it. This is not surprising. The human mind always tries to expunge the intolerable from memory, just as it tries to conceal it while current."

The flu was endured and quickly pushed from the collective consciousness despite the toll it had taken of humanity. Perhaps because the epidemic coincided with the relief that greeted the end of the slaughter of the war or maybe because it seemed to have taken its toll and then disappeared, once it was over, the epidemic was forgotten. Even the deaths of 400,000 Americans during the single month of October 1918 failed to create the panic that would accompany the annual polio epidemics that began to appear in 1916. Compared with the widespread carnage wrought by influenza, polio affected insignificant numbers of people, but many of them were children, crippled forever by what turned out to be a virus with an unusual habit. Although polio now seems like a bad dream from the past, when the gastrointestinal virus that causes it began its march through the population, it seemed novel as AIDS was in the early 1980s.

11

Gastrointestinal Viruses

Certain diseases leave behind them mystiques that remain in the collective mind for hundreds of years. Bubonic plague, which killed half the population of Europe in the mid-1300s, traveled from port to port with oceangoing rats whose fleas carried the bacterium *Yersinia pestis*. Called the Black Death, it is still the archetypal symbol of horror for all epidemics, described allegorically by Albert Camus in *The Plague*.

Richard Burton, the nineteenth-century explorer of East Africa, wrote that smallpox "sweeps at times like a storm of death over the land." A massive worldwide vaccination program ending in 1979 finally eliminated this viral infection, but many of its scarred survivors remain behind. Controversy continues over whether to destroy the last of the smallpox virus frozen in a few labs. The fear remains: Suppose the virus somehow got loose in what in another generation—except for Russian and American military personnel, who are still being vaccinated—will be an unexposed, unprotected population.

Although the 1918 influenza epidemic was unquestionably the most disastrous that has ever visited humankind, its fuzzy image failed to elicit the mythical trappings of less lethal diseases. Perhaps because polio killed and crippled children, it immediately took on

149

an aura of dread which was out of proportion to the numbers affected.

From the time epidemic polio appeared in 1916 through the 1950s, polio turned summers into battles between youngsters itching to do marvelous hot weather things and parents who were haunted by the dreadful vision of children and young adults dying or confined forever to iron lungs. Polio swept through the population year after year. One couldn't swim in a public pool, watch a parade, or even go to the movies. Anxious mothers hovered about making sure their children weren't getting tired or feverish and didn't have a cough or an upset stomach, all potential signs that the killer was at work.

Another factor that probably added to the near panic of the public was polio's novelty: The disease seems to have appeared in epidemic form only with the first outbreak of forty-four cases in Stockholm in 1887. Seven years later, a small town in Vermont recorded 119 cases. But in 1916, 6,000 Americans died and 27,000 were left paralyzed.

Polio is another viral infection, like Epstein-Barr virus (EBV), that appears to have become worse in tandem with improvements in sanitation. Babies born in the crowded conditions that often accompany poverty were exposed to the virus early and usually experienced only a mild cold or stomach upset, whereas the young of the well-to-do, protected from infection from birth, had no exposure and consequently no immunity. Pushing back the age of potential exposure put relatively privileged children and young adults at the highest risk.

Today, although vaccination has largely rid the western world of polio, 450,000 cases of the preventable infection occur yearly in less-developed nations. Polio is a relatively minor problem compared with the health toll of many other diseases in these countries and has been reduced to some extent by vaccinations, which World Health Organization officials estimate have been given to half the world's children.

The development of the polio vaccine by Jonas Salk in the mid-1950s caused a gradual reduction of the disease until by 1980 there were no naturally occurring cases in the United States. Although at first glance this seems to indicate universal protection, a substantial number of Americans remain unvaccinated, particularly poor children in urban areas. But even many of the unvaccinated are also protected because the oral vaccine that is most commonly used is made from attenuated rather than killed virus. This allows the virus to multiply in the gut for as long as three weeks after inoculation. During these weeks, immunized children shed the weakened pathogen and those with whom they come in close contact have a good chance of contracting it. These individuals in turn are also exposed to a nonvirulent virus and thus become immunized. This is called herd immunity, and it serves to actually crowd out more dangerous wild strains of virus from the environment. Expansion of the polio vaccination program in the Third World would have the same herd effect.

Despite and because of the vaccine, a few American children—perhaps as many as ten a year—pay a heavy price: Every so often, for unknown reasons, the attenuated virus reverts to its original virulence and causes the disease it was intended to prevent. These terrible accidents could be avoided by using the injectable vaccine made from killed virus, but the herd effect would be lost.

Other vaccines and the tissues in which they are grown occasionally cause similar dire accidents, and recently scientists at Columbia University and the State University of New York identified and cloned the gene that codes for the site used by the poliovirus to attach to cells. This is similar to the attempt to make a version of the CD4 cell receptor to which HIV binds; both decoys theoretically can attract and capture the viruses before they connect with cell surfaces. But in the case of polio, this innovative technique is far down the road and of low priority since an effective vaccine already exists.

Synergies

A contemporary account of the confusion, panic, and hype surrounding the development of a vaccine against polio was set down in loving detail by Greer Williams in his 1960 book *Virus Hunters*. In it, today's still-active elder statesmen of science Jonas Salk, Hilary Koprowski, and Albert Sabin (the first two now working to devise an AIDS vaccine) are depicted during the years when a very public debate raged over the relative merits—and safety—of a killed versus an attenuated vaccine. Financial gains and reputations hung in the balance, and vindictive personal attacks rivaled the acrimony that has accompanied research on AIDS.

The scientific advance that made it possible even to consider a vaccine for polio was the development of antibiotic drugs a few years earlier. In this case, it was the ability of penicillin to prevent bacteria from growing in test tubes that was essential. Just as the introduction of anesthetics allowed surgery to evolve from its dark age into today's often amazing refinement, the discovery of anti-biotics changed the face of laboratory science indirectly but very significantly.

Polios, which means "gray" in Greek, refers to the gray matter of the brain and nervous system, for which the polio virus sometimes has a deadly affinity. The overwhelming number of people who contract one of the three strains of the polio virus experience in-significant or no illness. But if, for reasons still unknown, the virus travels from the gut, which is usually the initial site of infection, to the brainstem, it causes bulbar poliomyelitis, which paralyzes the breathing apparatus: Without assisted ventilation, the patient dies outright. Attacks on the spinal cord sometimes result in paralysis of the limbs.

When vaccine development was first considered, the filterable agent that causes the disease had been found only in cells of the brain and spinal cord. Therefore, it was assumed that cells from those organs had to be manipulated to support the growth of the virus in test tubes.

In 1936, Albert Sabin and Peter Olitsky at the Rockefeller In-stitute saw that the virus refused to grow in cells other than those from embryonic brains. But this was a dangerous tissue to include in a vaccine. The consequences of injecting humans with substances created from nerve tissue were well known from the accidents in-curred when vaccines for rabies and yellow fever, which were man-ufactured in those tissues, caused a breakdown of the central nervous system of the recipients. This knowledge put the idea of a polio vaccine on the back burner for many years. It was only through the casual inclusion of some leftover polio-infected tissues in a cul-ture used in the search for a mumps vaccine that the virus's true nature was clarified.

Dr. John Enders's laboratory in the Children's Hospital in Boston in 1948 finally found a way to keep mumps alive in test tubes. This came about because the bacterial contamination that had thwarted long-term cultures of viruses in past years could be eliminated by adding penicillin. The antibiotic killed the bacteria but had no effect on viruses. In Enders's experiments this was of prime importance because, along with mumps, researchers were looking for a virus that was thought to cause diarrhea. This required that virus be grown in mouse intestines, which are always contaminated by bacteria.

During a series of experiments, several flasks of culture medium containing intestinal cells were left over. Instead of pouring the contents down the sink, on a hunch Enders took a test tube of polio-infected mouse brain from the freezer where it had been for several years and added it to the flasks. Without bacteria contaminating the experiments, the poliovirus infected, and continued to grow in, the mouse intestine.

Shortly thereafter the Boston researchers were able to prove that the polio virus initially infects cells of the stomach. This discovery was important enough for Enders and his colleagues, Fred Robbins and Thomas Weller, to be awarded the 1954 Nobel prize. Not only had they managed to grow the virus; more importantly, they had identified its target cell population. The way was open for the development of a vaccine, and scientific enthusiasm ran high.

At the same time Enders's lab began to grow the poliovirus, Dr. Gilbert Dalldorf, then director of the New York State Health Department in Albany, was investigating what appeared to be an outbreak of paralytic polio in children in the small town of Coxsackie, New York. Dalldorf was unable to find any traces of the poliovirus in the affected children but did find another, previously unknown virus. Widening their investigation to surrounding towns, Dalldorf's scientists found the new virus everywhere: Half the family members of the sick children had antibodies against it.

What they had in fact discovered was the coxsackie family of viruses, most of which we now know cause inapparent (unnoticed) infections. Today, some thirty members of this group are known, and many can cause a formidable array of illnesses, particularly in children. As well as affecting the heart, other strains cause rashes, oral ulcers, aseptic meningitis (with symptoms almost identical to those of other acute infections of the central nervous system), respiratory disease, and a paralysis that is almost indistinguishable from polio. Coxsackieviruses belong to the same family as the poliovirus, which they resemble in terms of size, resistance to killing, prevalence during summer months, and person-to-person spread.

Five years after their success in growing polio, Enders's group, still searching for a virus to account for dangerous infant diarrheas, examined many specimens from people with intestinal infections. Tissues from 300 children thought to have had paralytic polio turned up 72 who were infected with other unknown pathogens. These turned out to be among the first of a long list of what would come to be known as enteroviruses, which include members of the rotovirus, adenovirus, calicivirus, and coronavirus families.

Gastroenteritis

Worldwide, gastrointestinal (GI) illnesses cause half the deaths of children in developing countries. In the United States they are second only to the common cold in sending people to the doctor's office. Numerous viruses infect us through the stomach and cause GI upsets, but the Norwalk agent and rotoviruses are directly responsible for most diarrheal disease.

The Norwalk agent (probably a calicivirus) was first recognized in a town of that name in Ohio in 1968, where half the children in an elementary school had been laid low by "winter vomiting disease." Later outbreaks in summer camps implicated the water supply in distributing this highly infectious agent.

Over the years, studies have shown that outbreaks of Norwalk-caused stomach problems occur in the summer months, affect people of all ages, and cause symptoms for twenty-four to forty-eight hours. But the same studies have raised unanswered questions. Although most people have antibodies against the virus by the time they reach fifty years of age, resistance against reinfection seems highly capricious. Even volunteers previously infected with the virus who possess antibodies against it may or may not become ill when rechallenged with the agent. Strangely, those with the highest antibody levels tend to be more susceptible.

But the Norwalk agent is not the cause of infant deaths: That dubious honor goes to the rotovirus (Latin for "wheel," which they resemble) family. In the western world, rotoviruses cause acute diarrhea in infants and account for almost half of such infections requiring hospitalization. In the tropics, this agent is responsible for hundreds of thousands of deaths each year, largely from dehydration, of infants under two years old. Viruses belonging to the same family severely infect many young animals as well.

A vaccine against the rotovirus is desperately needed in developing nations, and many approaches to developing one are being studied. But a vaccine may not be possible: Antibodies against the rotoviruses confer only spotty immunity, which in any event lasts only a few years.

Like other GI infections, including polio, the causative pathogens of diarrhea can be spread through contaminated food and water supplies as well as through person-to-person contact. While polio causes paralysis and these other agents cause diarrhea and vomiting, there is yet another GI virus, the agent of hepatitis A, which primarily affects the liver.

Infectious Hepatitis

Forty to seventy percent of American adults have antibodies to the hepatitis A (HA) virus. This small, hearty DNA virus is spread like poliovirus and other enteric viral agents through contaminated feces, water, and food. HA infection that is contracted in early life is about 95 percent effective in producing protective antibodies. However, unprotected adults are at risk for illness of widely varying severity: One to three adults in a thousand die, while others may hardly notice that they've been infected. The usual illness consists of a week of chills, fever, nausea, and prostration followed by the yellowing of skin that liver damage causes. As the jaundice appears, despite looking ghastly, patients begin to feel well again.

Ninety percent of patients recover completely within three months, but a small number of people have a mild recurrence of liver disease within six months. The remaining 5 percent develop a chronic form of hepatitis. But hepatitis A is a piker in causing disease compared with its dangerous cousin hepatitis B. The discovery of this virus—the primary cause worldwide of liver cancer —by Baruch Blumberg is a shining example of how curiosity about one area of science may lead to the solution of a problem that hasn't even been posed.

12

Hepatitis B and the Cosmic Jigsaw

Important scientific discoveries seem to spring full-blown from the genius of a researcher and then make their way into headlines stating that a breakthrough has occurred. Most readers, if they think at all about the process that has led to the announcement, have a vague image of the "scientific method," which for most people involves a neat linear progression toward the "answer." After all, that's the way we dissected frogs and were told to combine chemicals in school: "If you do this and this, here's what will result." But by the time an experiment is reduced to a step-by-step procedure with a known outcome, it's already very old knowledge; the once-innovative experiment has been turned over to the lab technicians to do as part of their routine, the donkey work.

The linear, step-by-step method long used in the west has undoubtedly made possible the advances in science that have occurred over the centuries, but the time is coming when we are going to have to think in other ways about the interconnectedness of scientific enterprises. For instance, the rules that govern physics and the behavior of atoms are as relevant to how a virus maintains its envelope as they are to how weak and strong forces make the difference between the bonds holding together a sheet of steel and those that hold together a piece of toast. But the outpouring of scientific facts in the past 150 years is creating distances between scientific bodies of knowledge that badly need to communicate with each other.

Some call for a "general systems" approach, but a better way may be to come at science in an "other" way, perhaps in the way Oriental peoples have traditionally approached their paintings—with a roving perspective. In this mode, a mountain seems to have been painted after being looked at from ground level, from 100 feet in the air, and finally from the mountaintop. The result is a much different, perhaps more thorough "feeling" for the subject than that gained from the fixed, ego-oriented perspective of the west. Western sight lines lead with mathematical certainty from wherever the artist has chosen to stand, a linear approach which, while adequate for the steam engine, leaves much to be desired in today's science.

The more we learn about biological systems, the more they resemble a jigsaw puzzle of a painting done in the Oriental style. Some sections of the puzzle are pretty well assembled; for instance, most of us understand that the world is more or less round and that the mating of a maid and a bull won't produce a Minotaur. However, the more we know, the more fantastic become the possibilities of what there is to know; in these days of genetic engineering and other extraordinary technologies, few scientists are incautious enough to attempt to glue any one piece of the puzzle permanently to another.

Having chosen which section of the puzzle to work on—such as a particular type of cell or a category of virus—a scientist will find the bits scattered before him or her to be turned this and that way, tried and rejected, and sometimes even forced together despite their lack of fit. The trickiest part of the assembly is that there are no corner pieces or smooth edges to guide the puzzle solver and no final shape, never a time when one can step back and gaze in satisfaction at a completed picture. At the next table and the next hundred thousand tables, other scientists are assembling sections which in one way or another will ultimately have to link up with everyone else's puzzle. If there's an ultimate design, a final "the way it is," it's out there in a distant, inconceivable future.

Physicists are convinced that a unified theory of everything, which explains how the universe came into being, is within their grasp. Biologists aren't nearly so sanguine because what they're working with is the process of life itself.

Since no one knows what the puzzle is supposed to look like, fitting in one of the important pieces seems to require luck combined with the "prepared mind" haughtily described by Louis Pasteur in simpler times.

Luck often depends on the time at which an idea or intuition surfaces. A relevant case in point is that of Dr. McKinney, the plant pathologist who discovered in the 1930s that viruses can mutate. Because the idea of viruses was so inchoate and the ability to look at them was equally undeveloped, McKinney "was ridiculed for twenty years" and died unnoted, as Dr. Theodore Diener recalls. McKinney's fate had been to come up with an elegant realization at a time when neither the equipment nor the intellectual climate was right to prove his thesis. Being ahead of one's time is probably more painful and certainly more frustrating than being behind it.

A contemporary fable that supposedly demonstrates luck in science has been created from Sir Alexander Fleming's discovery of penicillin in 1928. Schoolchildren are taught that a mold spore drifted in through the open window of his laboratory, settled on a culture, and killed the bacteria growing there. Wasn't he lucky to have noticed that in his cluttered lab? But while the details are accurate enough, the reality of Fleming's discovery of the antibiotic effect of *Penicillium* fans out in time both before and after his observation.

Surgeon Joseph Lister had made the same observation in 1871; English microscopist John Tyndall, Louis Pasteur, and French physiologist Paul Vuillemin had each noted the effects of the *Penicillium* family of molds on bacteria around the same time. All these researchers were intrigued with the mold's potential usefulness in human infections and all published papers suggesting the importance of the substance, but the idea went nowhere. The only scientist of the time who had used *Penicillium* in actual animal experiments was a young French physician, Ernest Duchesne, who died in his thirties, leaving behind a thesis that the mold should be looked at with an eye to human therapy.

Fleming scarcely went further himself. It took the emergency medical situations created by World War II to mobilize government resources plus the expertise of the growing pharmaceutical industry to turn the observation into the fact of a drug. In effect, by the early 1940s there were enough pieces of the jigsaw on that particular table to fit together a whole section. The result was penicillin, the first man-made antibiotic.

When the ultimate reward in science, the Nobel prize, was given in 1945 for physiology, Fleming shared it with Howard Florey and Ernst Chain of Oxford, the researchers who had finally isolated and purified the active ingredient in the mold and then proved its therapeutic effects. The luck of recognizing that the pieces fit together required that the times—and all that that implies—be right.

Hepatitis B: Blumberg Goes to the World of Nature

When Baruch Blumberg set out to identify the differences in inherited immunologic traits in human populations, he literally fell over the virus that causes hepatitis B.

"It is clear that I could not have planned the investigation at its beginning to find the cause of hepatitis B," an experience, Blumberg says, that "does not encourage an approach to basic research which is based exclusively on specific-goal-directed programs for the solution of biological problems." Not, in other words, the western approach to problems.

Blumberg's discovery shows where luck in science diverges from the kind of luck involved in pulling the lever on a slot machine or being dealt a good hand of cards. By remaining ready to fit any new piece into the puzzle, no matter where it might go, Blumberg saw a pattern it hadn't occurred to him even to look for. His observation, which gained him the 1976 Nobel prize for biology, has already saved many thousands from sickness and death and in the future will save millions more.

Always a man with wide-ranging interests, Blumberg was a commander of landing craft during World War II, held a ticket as a ship's surgeon, and worked as a merchant seaman and occasionally as a hand on sailing vessels. He might have pursued physics after obtaining his undergraduate degree in that subject, or he could have followed up on his graduate work in mathematics. Conceivably, Blumberg could have concentrated on medical anthropology, which, along with medicine, he has taught at the University of Pennsylvania, or he might have gone into epidemiology, one of his favorite areas of medicine.

All these interests are hinted at by the eclectic collections of books and memorabilia that fill Blumberg's office in the quiet enclave of the Institute for Cancer Research outside Philadelphia. Framed by books detailing obscure folk medicine and botany, he speaks with frustration of the difficulty of obtaining funding for research into plants with therapeutic value, 1,200 of which his team at Fox Chase Center has already tested.

While Blumberg's potential for almost any field seems unlimited, at his father's urging he decided on medicine when he was a young man. Working toward his M.D. at Columbia University, from which he graduated in 1951, Blumberg spent several months in the remote upcountry areas of northern Suriname, delivering babies and participating in the first malaria survey done in the region.

There he came across the important idea that would set him on his future scientific track: "the enormous variation in the response to infection with *Wuchereria bancrofti*," the parasite that causes elephantiasis. The laborers on the surrounding sugar plantations included native Indians, Hindus, Javanese, Africans, Chinese, "and a smattering of Jews descended from the seventeenth-century migrants to the country from Brazil."

Returning home to finish his training at New York City's Bellevue Hospital, which he describes as "reminiscent of Hogarth's woodcuts of the public institutions of eighteenth-century London," the young doctor established his lifelong concern for human beings as the ultimate beneficiaries of esoteric biomedical research.

"Anyone who has been immersed in the world of a busy city hospital, a world of wretched lives, of hope destroyed by devastating illness, cannot easily forget that an objective of biomedical research is, in the end, the prevention and cure of disease," Blumberg said in the preface to his Nobel address.

Instead of plunging immediately into the practice of medicine, Blumberg carried his interest in a chemical called hyaluronic acid to Balliol College at Oxford in England. There he wrote his Ph.D. thesis on this sugary substance that binds and protects connective tissues and is intimately associated with the various arthritislike diseases and hence in disorders of the immune system. Integrating this with his earlier curiosity about different racial resistances to disease, Blumberg focused on *polymorphism* (*poly* means "many"; *morph* means "form"), which is the ability of different types of individuals to exist in the same group or species.

With no thought of hepatitis in mind, Blumberg began exploring the reasons different people, particularly those belonging to various racial groups, appear to respond differently to infectious agents. Just as differences in the genetic makeup of the individual account for different blood groups, Blumberg hypothesized that more subtle variations might be responsible for the variety of ways in which individuals respond to the same infective agent.

The many, mostly incompatible, red blood cell types such as O, A, B, and AB that occur in the single species of human beings is the most familiar example of polymorphism. Another important blood group difference is the Rh factor that may result in an Rh-negative mother aborting an Rh-positive fetus. It isn't blood per se that's polymorphic; rather, it's that the various components of blood are coded by specific genetic markers—our individual pedigrees—and have to pass muster by the immune system. These differences explain why in blood transfusions the donor's and recipient's blood types have to be matched within various groups to avoid rejection. The same holds true in organ and bone marrow transplants but not in transplants of the cornea; because the cornea has no blood vessels to carry immune cells, it escapes surveillance.

The goal of Blumberg's research was to find out what impact inherited differences have on the susceptibility of various groups with a specific set of genetic markers. He began by looking at patients who had had multiple transfusions. Would they develop antibodies against proteins that they hadn't inherited but that were present in the donors' blood? Or would they learn to tolerate these foreign antigens?

Polymorphism and Host Defenses

Only people produced from the division of a single egg and sperm—identical twins—share the same genetic markers. The rest of us exist quite alone in a biological sense.

"Polymorphic antigens [those that cause an immune response]

may have an effect when one human's tissues interact with another's in blood transfusions, transplantations, pregnancy, intercourse, and possibly . . . when human antigens are carried by infectious agents," Blumberg writes.

Differences in the genetic codes of individuals are certain to play a major role in host defense and in one's unique response to illness. Many of these differences can be read by analyzing a drop of blood or serum.

While at the National Institutes of Health from 1957 to 1964, Blumberg began collecting blood products. By the time he won the Nobel prize, his group, then at Fox Chase in Philadelphia, had accumulated more than 200,000 specimens. Frozen away were the genetic fingerprints of most of the world's racial groups as well as those of many animals, including even the zebu cattle belonging to the Fulani people of northern Nigeria. How were people different? How the same? And what difference did it make in regard to infectious diseases?

An ingenious technique called gel electrophoresis was devised in 1955 to separate blood proteins on the basis of complex characteristics of size and shape. With this method, an individual genetic pattern reads out as vertical columns of smudges with distinct patterns, a sort of vertical fingerprint. This technique is used to determine whether a particular man can or cannot be the father of a child, and more recently it has been used in court to prove or disprove the guilt of a suspected rapist.

Blumberg and his colleagues took advantage of the new technology to look at the genetic makeup of persons from Africa, Europe, and Alaska. They "found striking variations in gene frequencies."

During the next few years the team studied Greeks, Micronesians, American Indians, and blacks and whites, in whom "the richness and variety of biochemical and antigenic variation in serum became strikingly apparent." This raised many questions that could at least be posed: The one Blumberg selected was, Do patients who receive frequent transfusions develop antibodies against one or more of the foreign proteins present in the matched (by blood type) but by no means identical blood of donors? If so, could sera against those antibodies be used to identify the many different proteins of the individual? In effect, Blumberg was searching for a way to accomplish what is now being done with targeted antibodies, which are used as probes to spot specific molecules.

Blumberg recalls the day in 1963 when, after testing sera from thirteen patients who had had multiple blood transfusions, they found a completely novel protein band in the gel. "It was a very exciting experience to see those precipitin bands [indicating proteins] and realize that our prediction had been fulfilled," Blumberg recalls. The blood under scrutiny was that of a hemophiliac.

Hemophilia is an inherited trait that is usually passed on from a healthy mother to her sons, who, because of their inability to produce a certain protein in the blood—clotting factor VIII—have extremely prolonged coagulation times. The information for blood clotting is carried on the X chromosome; because women inherit two X chromosomes, few suffer from severe hemophilia, since the good X is able to overwhelm the effects of the carrier X. By contrast, men inherit the father's Y and the mother's X. If the X is programmed for the clotting defect, the boy will be a hemophiliac. This is a different kind of genetic inheritance from that of sickle cell anemia, which, when carried by and passed on from one parent, results in the offspring having the sickle cell trait but not the disease. For sickle cell disease to occur, both parents must have and pass on the genetic trait.

Aside from the well-known danger of bleeding to death from a minor wound, the hemophiliac's inability to control leakage from tiny blood vessels after even a minor bump or bruise causes a range of disabling problems to the joints and other body systems. To prevent this, factor VIII must be replaced on a regular basis; until recently, when a genetically engineered factor VIII became available, it took the combined blood of hundreds of donors to gather the needed protein. Since each factor VIII injection represents many blood donors, hemophiliacs were the logical group in which Blumberg could evaluate the effects of multiple transfusions.

Tested against a range of other blood, the blood of the New York hemophiliac reacted only with that of an Australian aborigine. The significance of a novel protein that was shared by a New Yorker and an aborigine was a mystery. Blumberg called it Au after Australia, and his team began looking at other blood samples to see if the novel Au antigen was present in any other populations.

The Surprise Finding of the Cause of Hepatitis B

Unlike diseases such as polio and smallpox, which attract a lot of attention, hepatitis is just a couple of dull-sounding viral infections in which the public has never been terribly interested. But on a worldwide scale, the several known types of these viral illnesses are serious indeed.

As a chronic infection, hepatitis B affects 5 percent of the world's population. In areas such as China, southeast Asia, and tropical Africa, up to 70 to 95 percent have been infected, and 15 percent of these people may be healthy chronic carriers who are able to pass the virus to those with whom they have close contact.

Hepato is the Greek word for "liver," and medical terms starting with this prefix refer to that organ. *Itis* means "inflammation," so although various types of hepatitis are caused by different agents, all are designated by their inflammatory effects on the liver. Regardless of the eventual outcome of different types of hepatitis, they all start by affecting the liver, which becomes tender and swollen and causes jaundice from the release of bilirubin. This product of the breakdown of aged red blood cells in the liver discolors the whites of the eyes and skin, creating a decidedly tobacco-stain yellow cast, a sign that the liver isn't cleaning the blood properly.

Hepatitis A, which is an acute infection of the liver, is caused by a virus that is spread mostly in feces-contaminated food and water, and is quite contagious (Chapter 11). Another type of hepatitis, which has been called non-A, non-B (NANB), has very recently been identified as being caused by yet a third virus and is now tentatively called hepatitis C. These three forms of infectious hepatitis have similar symptoms, such as a loss of appetite, weakness, and jaundice, but after a few weeks most people recover fully and are thereafter immune against them.

On the other hand, chronic hepatitis B virus (HBV) infection—chronic because the immune system hasn't been able to eliminate it—is the most important cause of cancer of the liver. This is a case where a virus transcends what we usually think of as infection and takes on the sinister role of a cancer-causing agent. Where HBV is endemic in underdeveloped parts of the world, liver cancer probably accounts for as much as 30 percent of all cancers. It's been estimated that 5 percent of the world population is infected by HBV, including a million Americans. In Europe and the United States about 5 new cases of liver cancer per 100,000 people occur each year, and the incidence is rising. The World Health Organization maintains that HBV is second only to cigarettes as a proven cancer-causing agent.

The last type of hepatitis is delta hepatitis. This disease (Chapter 13) apparently is caused by a defective viroidlike fragment that is remarkably similar to the rolling circle discovered by Theodore Diener in potato plants.

The Difference That Differences Make

Blumberg had started with the theory that polymorphism (or basic genetic differences) plays a major part in the health of humankind and designed a research approach he thought might prove it. In two senses he got extremely lucky. One piece of luck obviously was the discovery of HBV; the other, as will be seen later, was that the virus itself became the cornerstone for exploring his original theory.

After finding the Au antigen in the New York hemophiliac, which the researchers still thought represented a novel individual genetic marker, the team began testing hundreds of samples of stored blood. They found that the newly discovered protein was rare in healthy Americans but was quite common in the blood of persons from the Philippines and in certain Pacific populations. If, as they then assumed, Au was an inherited marker of certain racial groups, this difference would make good sense.

But then the team took another tack and began looking at blood samples from leukemia patients at the University of Pennsylvania's medical center. The samples were very often found to be Au-positive. Did the protein have some relationship with cancer?

To explore this question, the team examined the blood of a group of children known to have an exceptionally high likelihood of developing leukemia—from 20 to 2,000 times that of other children—those with Down syndrome (mongolism). Thirty percent of the blood from Down patients living in an institution near Philadelphia was positive for Au.

"Until this time all the individuals with Au who had been identified either lived in Australia or some other distant place or were sick with leukemia," Blumberg says. But in what possible way could patients with leukemia or Down syndrome genetically resemble aborigines?

"Scientific research has an infinite quality," says Blumberg. "The more we know, the more we know about what we do not know, and this unknown in turn must be understood."

Then one Down patient who had tested negative for Au was found positive on a subsequent blood test. One mystery thus began to unravel.

"Because he apparently had developed a 'new' protein [rather than having been born with it, as would have been the case in Blumberg's original theory], and since many proteins are produced in the liver, we did a series of 'liver chemistry' tests," recalls Blumberg.

The tests showed that the patient, James, had developed a form of chronic hepatitis.

This raised a completely different possibility. Au could not be, as they had thought initially, "an inherent characteristic of the individual" but instead must be associated with hepatitis. An accident in the Fox Chase laboratory confirmed it.

One day in 1966 Dr. Barbara Werner, who had been working on Au samples in the lab, realized that she had been feeling less well than usual. Aware of the possible connection between the new protein and hepatitis, she examined her own blood.

"The following morning a faint but distinct line [on the gel test] appeared, the first case of viral hepatitis diagnosed by the Au test," Blumberg related in his Nobel address. "Diagnosed by," but what was it? A marker of viral infection? A protein often found in association with hepatitis? The tracks of the virus itself? The latter possibility seemed very likely, so the case and speculations about what it might mean were written up and sent to a medical journal.

Explains Blumberg, "Our original publication [suggesting that the Au protein might be the cause of hepatitis] did not elicit wide acceptance; there had been many previous reports of the identification of the causative agent, and our claims were naturally greeted with caution." Even a later paper expanding on research results was rejected on the grounds that it was simply "another candidate virus," of which a gracious plenty had erroneously been reported over the years as being the cause of hepatitis B.

Convincing evidence that Au was probably the cause of hepatitis B began to accumulate. While the Blumberg group, in cooperation with the University of Pennsylvania, was designing a long-term study of transfusion recipients, researchers in Japan moved ahead. Professor Kazuo Okochi, one of the many researchers with whom Blumberg had shared his testing material, found that Au could indeed be transmitted by transfusion and that those who received blood from Au-positive donors often developed hepatitis.

The Philadelphia study was immediately aborted, which Blumberg describes as "a dramatic example of how technical information may completely change an ethical problem. As soon as the results of Okochi's well-controlled studies became available to us, it became untenable to administer donor blood containing Australian antigen. *Autre temps, autre moeurs.*" By the fall of 1967, all donor blood from hospitals working with the Fox Chase group was tested to exclude Au-positive donors.

A positive change was immediately apparent: The number of blood recipients who contracted posttransfusion hepatitis B fell by two-thirds. Within a year of the original discovery, the research project was put into practical application in many hospital blood banks, which began discarding HBV-infected donations.

Though this sounds like an eminently sensible thing to do, such drastic changes in medical practice are seldom made this quickly or, as Blumberg describes it, on the basis of largely "nonrational factors." "Nonrational factors" refers to a story that ran in *The New York Times* in July 1970 stating that the cause of hepatitis B had been found and that the virus was being transmitted in blood transfusions. Subsequent to the article, several successful lawsuits were brought against hospitals and blood banks for transfusing Au-positive blood. Courts ruled in favor of the plaintiffs in several instances, and this doubtless helped speed the change in policy.

In rapid succession New York and several other states passed legislation mandating Au testing; the American Association of Blood Banks followed, and by 1973 testing was required in all blood banks. The annual saving in preventing transfusion-caused hepatitis B in the United States alone is estimated to be in excess of half a billion dollars.

Shortly thereafter, Au was shown to be the surface antigen (outer coat) of HBV. It took more than fifteen years from the initial

Au finding to actually see a virus: Not until 1986 was a way found to grow HBV in the test tube. The virus itself is a sausage-shaped entity called the Dane particle. Even before the virus was ever seen, a vaccine against it was made by extracting the Au protein, against which the original vaccine was made.

One of the great contemporary medical researchers, Nobel laureate Sir Peter Medawar, wrote in the 1960s, "Scientists are building explanatory structures, telling stories which are scrupulously tested to see if they are stories about real life." Despite the importance of the discovery of the Au antigen, for Blumberg it was a subplot that in no way distracted him from his major theme: the search for underlying genetic differences and their implications in regard to susceptibility to disease.

Describing the approach that led to discovery of the Australian antigen, Blumberg told the 1976 Nobel audience in Stockholm that "the genetic hypothesis has proven to be very useful not in the sense that it is necessarily 'true,' but because it has generated many interesting studies on the family distribution of responses to infection with hepatitis B." Though hepatitis itself may attract little media attention, the virus apparently has a great deal to do with determining the sex of children born to infected parents.

HBV Also Determines Sex

Vive la différence in the context of health and longevity can be said with a great deal more conviction by women than by men, particularly when it comes to congenital diseases. It has been proposed that the gene controlling the production of antibodies is on the X chromosome, giving women a greater variety of possible responses to infective agents than men have. However, this efficiency of immune response may account for some of the diseases thought to involve hyperactive, inaccurate immune responses, such as lupus and rheumatoid arthritis, that affect women eight times more frequently than they affect men.

But Down syndrome and hemophilia, as well as severe ankylosing spondylitis (an autoimmune disease of the spine), affect young men almost exclusively. Men are also more likely to contract leprosy and certain kinds of cancers, such as Hodgkin's disease and myelogenous leukemia. A study from Johns Hopkins that followed patients for sixty-five years showed a higher frequency of bacterial meningitis and septicemia and their associated mortality in male children.

Once infected with HBV, men have almost a 70 percent chance of developing the chronic infection and suffering its potentially severe outcomes, while women have only around a 33 percent chance. How a virus is able to preferentially infect or cause more severe illness in one sex than another is hard to fathom, but it must have something to do with the different sex hormones: androgens in males and estrogens in females.

Was it possible, Blumberg wondered, that some cross-reactivity (similarity) between sex-associated antigens (markers on cells) and Au is responsible for the different course of hepatitis in men and women? Viruses are thought to play a role in kicking off the autoimmune process in some people when there is a similarity between markers on the host cell (determined by the individual's genetic pattern) and markers on the virus.

Humans infected with mumps, measles, herpes, hepatitis, and other viruses have been found to develop antibodies against their own proteins: This is called molecular mimicry. Moreover, HBV is a poor sort of virus that has to assemble a new coat from the proteins it pirates from infected cells. This creates a high likelihood of the immune system recognizing viral offspring—wrapped in "self"—as part of the body's own tissues and therefore ignoring it.

An even stranger apparent effect of HBV that Blumberg wanted to pursue was how the virus might play a role in the numbers of male or female babies born in the first place.

Nature dictates that humans produce about 106 male babies for every 100 females. Although female-programmed zygotes (fertilized eggs) may die in larger numbers early in conception, evidence from stillbirths and spontaneous abortions and postnatal death rates show that male mortality is considerably higher than female. This works out so that by puberty the sexes are pretty evenly matched in terms of number.

Based on the knowledge of these differences, and particularly on the different rates of HBV incidence in men and women, Blumberg's group developed the hypothesis that if HBV was similar to a normal male-associated antigen, it might slip past the immune system without being recognized. It would follow that female immune systems, lacking the male-coded protein, would be more likely to recognize and launch an antibody attack to eliminate HBV.

Blumberg followed the premise that there is an inherent protein in HBV that mimics an antigen associated with a male antigen. His proposition was that the immune system of a woman who had been exposed to HBV and had developed antibodies against it could, when the woman was impregnated by a sperm carrying the Y chromosome, "react with male antigens [of the fertilized egg] and perhaps hinder fertilization by sperm bearing a Y chromosome or increase the probability of spontaneous abortion of male fetuses."

To examine his theory, Blumberg looked at a Greek village with a high infection rate of hepatitis B. An extensive study showed that couples had the culturally usual number of sons but had fewer daughters than expected. As is true in many cultures, the birth of sons is considered far more desirable than that of daughters. Blumberg hypothesized that when both parents have hepatitis but not the antibodies to attack a male fetus, "they have the desired number of sons early in marriage and, as a result, restrict the number of subsequent births, and therefore the total number of daughters born."

"Alternately," Blumberg says, "HBV transmitted from a carrier parent [with *no* antibodies] might replicate more rapidly and be more lethal to female than male embryos, while mothers with antibodies would protect female embryos from HBV infection, thus eliminating selection pressure on female fetuses."

"Human behavior is affected by hepatitis B infection," Blumberg states, which splendidly fulfills its role as an agent of change.

The Iron Connection

Another major difference in the outcome of hepatitis B is predicated on the age at which one is infected. The chance of becoming a carrier who later develops liver cancer is greatest when a person is infected by HBV in infancy, because the cancer takes many decades to develop. Blumberg said almost in wonderment that "when I realized that we didn't have to *cure* HBV but could delay it and thereby prevent liver cancer, it was a real breakthrough, an important idea." Curing a persistent viral infection is much like trying to cure cancer. Since normal cells have the same metabolic needs that infected or cancerous cells have, an agent that kills one kills the other: Cancer therapy is limited by this panoramic destruction of healthy cells.

Searching for a way to at least retard the gradual evolution from a merely infected to a cancerous liver, Blumberg's group came across another part of the puzzle in an article from Holland. Dutch researchers working with kidney failure patients who undergo frequent dialysis (the mechanical cleansing of blood) reported that the frequent HBV infections experienced by their patients were often associated with a rise in hemoglobin, or iron levels in the red blood cells.

Iron, which plays an essential role in carrying oxygen throughout the body (an iron-deficient person is usually anemic), also wears other hats. For one thing, it is required by bacteria, fungi, protozoa, and tumor cells for their growth. The internal heat caused by the fever that accompanies an infection helps keep bacteria from multiplying by preventing the germ from using the host's iron stores.

Many trade-offs involving too much and too little iron are known to exist: People genetically programmed to produce abundant iron often develop cancer; those with active tuberculosis, which depletes iron, have fewer cancers. Alcohol increases iron levels; malnutrition lowers them. These issues have not yet been included in studies of causation.

While moderate restriction of iron, which we ingest in our food, may enhance the immune system and make it better able to fight infection, too little has the reverse effect. A dramatic example of this is seen in refeeding camps for African victims of famine. Sudden improvement in nutrition, which includes high iron supplements, causes an increase in the frequency and severity of infections, including malaria and tuberculosis. In HBV infection, patients destined to become carriers routinely have higher iron levels.

"We made the hypothesis that increased iron stores might increase the possibility of the development of primary hepatocellular carcinoma in carriers, that is, that iron is involved in the pathogenesis of this cancer, and that alterations in iron or iron storage could be used to favorably alter the outcome," Blumberg says. He hopes that finding ways to manipulate the amount of iron available to infected liver cells can be used to slow cancer development, allowing

HBV-infected people to live out their normal span. Blumberg calls this "prevention by delay."

When this is combined with prevention through blood screening and use of the new HBV vaccine (the first ever made by genetic engineering) to prevent early infection, one of the world's major infections—and the leading cause of an untreatable cancer—may some day be a thing of the past.

Blumberg equates the scientific process with the myth of Daedalus, "who was a legendary figure of the classical world, a craftsman, inventor, architect, and artist. He was also a generator and solver of problems. Every time a question was answered, it raised several more, and these in turn led to other questions; his search was endless." The discovery of the cause of hepatitis B raised numerous social problems surrounding transmission, but it also came in time to define the evolving recognition that an additive companion virus was causing an acute and even more dangerous type—delta hepatitis.

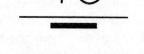

13

The Delta Fraction

Some Fluky Thing

Tall, lean John Gerin wraps his feet around the chair legs, snuffling with amusement. "It was such a long shot. We were just at the beginning of our knowledge of this agent and its biology. You know that when the only way to define an entity is by describing what it *isn't*, you're a long way from knowing what it is," he admits.

Even the existence of the unknown agent Gerin refers to was unsuspected until Mario Rizzetto, a young physician in Turin, Italy, saw something in the blood of a patient with hepatitis B "that didn't make sense."

"How many times are things thrown in the closet?" Gerin asks rhetorically. "God knows!"

Gerin's preoccupation with hepatitis began in 1967, shortly after Baruch Blumberg discovered the Australia antigen in patients with hepatitis B. As head of the division of molecular biology and immunology at Georgetown University's research station in Maryland, Gerin took a chance that Rizzetto's strange finding might turn out to be important.

Although Rizzetto had published his discovery of the novel protein, "the European community was terribly unimpressed," Gerin recalls, a situation which duplicated the experience Blumberg had had with his first reports of a hepatitis antigen because "everybody was finding antibody and antigen" that was thought to be involved in hepatitis. Indeed, many of these discoveries turned out to be merely chemical noise; others were autoimmune complexes, which were hardly novel substances.

Confronted by lack of interest at home, the young Italian wrote Robert Purcell at the National Institutes of Health, inquiring about the possibility of working with American investigators.

"Bob sent it on to me because it was more in my line," Gerin recalls. "I thought it looked pretty good and decided to bring him over here. We got a lot of negative feedback, but we asked him over anyway."

Rizzetto arrived with tissues and blood sera and, with the Georgetown group, began working on a method they hoped would tease out the new substance, if indeed there was one. The initial experience was disastrous, Gerin says.

"To get that antigen out of that liver took almost nine months. It's a very unusual protein."

However, even a test for the putative new substance was merely a first, tentative step. Could it be found in persons with hepatitis B or other liver diseases? Was it transmissible? The researchers couldn't rule out the possibility that Rizzetto's finding was just a previously undetected mutant of hepatitis B. Such mutants have been found recently and may account for as many as a third of cases that slip through blood screening for hepatitis B virus (HBV).

The real test of whether the protein was a transmissible agent was to study its effects in a chimpanzee already infected with hepatitis B. Ordinarily, HBV doesn't cause nearly as severe a disease in this animal as it does in a person. When injected with the new protein, the chimp suddenly developed acute hepatitis, but even this failed to disprove the presence of a mutant variant of HBV.

An early difficulty the researchers ran into in trying to determine what was actually going on with the sick animal was that the markers indicating HBV infection seemed to almost disappear after injections of sera containing Rizzetto's agent were given to the chimp. It was this disappearing act, in fact, that proved that a virus other than HBV was active. "This showed for the first time that it was a separate entity," Gerin explains. Whatever the new agent was, it had the peculiar ability to dampen the activity of HBV.

Finally, in 1977, Gerin's team found an agent unlike any known to infect humans and one which proved to be the unique and important viral agent that causes the most severe type of hepatitis—the delta particle.

"The finding of some fluky thing like that, which probably would have just been discarded, just something left in the literature," was accomplished by a stubborn determination which Gerin characterizes as "a fun time, a good experience. You always figure that there's nothing new under the sun. But not in this case; there just wasn't anything like this." There was, but in potatoes, not animals; the agent was suspiciously similar to the one found by Theodore Diener.

Return to the Potato

Like Theodore Diener's potato viroid (Chapter 3), the delta particle is a small circular molecule of RNA that is remarkably similar in size and habit to the satellite viruses of plants. Where did it come from? Was it a case of convergent evolution along with viruses, similar to the way various animals evolved on different continents —alike but not the same? Or perhaps the delta particle represents a snipped-out sequence from another virus, an intron that became independent. Perhaps, suggests Diener, the delta agent is an evolutionary leftover, a living fossil. "RNA molecules that have survived and evolved since their origin during the very early prebiotic stages of evolution" is how Diener describes this possibility.

Like the potato viroid, these strange particles are the ultimate parasites, completely dependent on the ability of the host's cells to replicate. Even the simplest viruses are equipped with genes that code for the new envelopes that protect their progeny's genetic material, but not viroids, and not the delta hepatitis agent.

"If you look at the forms," suggests Gerin, "it appears to replicate by a rolling circle model and may be capable of autonomous replication.

"It's fun to draw parallels and draw hypotheses, and in this case they allow you to draw testable hypotheses; because if you track the way the viroid people are doing their experiments, it benefits us all," he explains.

Of the origin of delta, Gerin is inclined toward the idea that

"RNA was the first molecule and delta is probably that kind of mechanism. It probably needs a host enzyme for the first round of replication" and then "tricks a normal enzyme into thinking it's a template; then it goes through self-catalyzed reactions." The self-splicing, rolling circle cuts, sorts, and reattaches itself. As in introns, the noncoding part of a virus is cut out and discarded during translation from one form of genetic material to another.

The Nice Little Critters

Delta is strange in ways that go beyond its close relationship to plant viroids.

"You may have a virus that doesn't respect species anymore," muses Gerin.

At Fox Chase other examples of HBV that infect several species of animals have been discovered.

"HBV is the prototype of a whole family of viruses in other species, which is quite unusual." One example of these animals are woodchucks, which Gerin refers to as "nice little critters; we just pick them up on the street."

This is a neat way to obtain lab animals, given that 30 percent of woodchucks are already chronically infected with an HBV-like virus, which they transmit only to others of their species. "We asked the question: Would the woodchuck hepatitis support the replication of delta?"

In a series of experiments, the researchers found that hepatitis-infected woodchucks can also contract delta.

"In this case, the delta can get into the hepatocyte [liver cell], so there must be some receptor [for it] there," Gerin states.

In replicating, the delta virus picks up its envelope molecules from the woodchuck's own cells, so it can be easily identified as having been produced in the animal's body rather than being merely the agent injected into the animal.

Next, the researchers injected HBV-infected chimpanzees with the woodchuck-coated delta particle. "The chimp came crashing down with delta hepatitis," Gerin reports. The virus they recovered was delta. It had lost its woodchuck coat and picked up a chimpanzee molecule coat.

Few viruses readily cross from one species of animal to another, and woodchuck to chimpanzee is certainly a quantum leap insofar as different species go. Gerin suspects that the delta particle has its own receptor, which is programmed for the liver cells of every species, but only in the presence of HBV. Eliminate HBV, and delta appears to disappear, too.

Delta hepatitis is a unique disease made possible by cooperative interaction between two pathogens that have no relationship to each other. HBV is a DNA virus, while delta contains only RNA.

However, HBV has an extremely peculiar habit for a DNA virus: Uniquely, it must undergo a step otherwise known only in retroviruses—the use of the reverse transcriptase molecule—to replicate itself. Perhaps this accounts for the interface between HBV and the RNA of the delta agent, but no one knows for sure.

However, chronic carriers of HBV may eventually benefit from drugs designed to interrupt the reverse transcriptase step of the AIDS virus. Hiroki Mitsuya of the National Institutes of Health is testing nucleoside analog drugs similar to azathioprine (AZT) in Peking ducks, which, like woodchucks, contract naturally occurring hepatitis.

Investigating HBV by using the duck disease (until late 1986 HBV couldn't be grown in the laboratory), which he calls "a pretty good model for hepatitis B," Mitsuya has found that all the nucleoside antivirals "cause a significant decrease in DNA polymerase [reverse transcriptase] activity and hepatic viral DNA." It is possible that AIDS research will do for HBV what it is doing for cancer research and for our understanding of the immune system and even provide a therapy for chronic HBV carriers.

But the discovery of the delta fraction also may hold an answer to the question of therapy through its ability to shut down the activities of HBV. Researchers are actively searching for something that will thwart this double infection, because when delta adds its weight to HBV, a dire illness follows.

The Answers That Beget More Questions

Delta hepatitis can cause either acute or chronic infection depending on whether it is contracted at the same time as hepatitis

B—a process called coinfection—or contracted after a person has developed chronic hepatitis B—a process called superinfection. Coinfection causes an extremely acute illness with a mortality rate of 2 to 20 percent. (This contrasts with acute hepatitis B alone, which has less than a 1 percent death rate.) However, when antibodies against HBV finally dispense with that virus, delta also disappears.

Superinfection is far more serious. Once HBV has established a chronic infection, 70 to 80 percent of patients who are subsequently infected with delta develop cirrhosis of the liver, as opposed to the 15 to 30 percent whose livers are damaged in this manner by HBV alone. Moreover, 15 percent of these people develop fulminating, fatal cirrhosis within two years.

In acute HBV-delta infections, even antibody may disappear so that no signs of either virus are left behind after recovery. It's possible, though, says Gerin "that delta is waiting there for another HBV infection." The dual activity of these viruses is typical of some plant satellite viruses, and the Georgetown researchers are putting delta in animal models, waiting, and then challenging them with HBV. One of the big problems is that delta can't be grown in tissue cultures, but the new ability to grow HBV in the laboratory should speed the work along.

An intriguing possibility raised by the Georgetown researchers is that because delta "shuts down HBV," it may be possible to use delta to therapeutic advantage. On its own, of course, delta is too dangerous unless, Gerin speculates, "you could build an attenuated delta virus," which basically would function as a "stop signal" for HBV.

"If we can identify the [gene] sequence causing pathogenicity," says Gerin, "we could build a delta that could shut off B and not cause disease." Infection with delta also quiets infection with the mysterious hepatitis C agent, although the two apparently share no similarities except their common success in failing to elicit an effective immune response.

The inadequacy of the immune response in people in whom hepatitis becomes chronic is a miniquandary within the larger problem. Italian researchers have found that antibodies made against a part of the delta particle also act against the thymus gland, the nut-sized clump of cells lying behind the breastbone which are responsible for turning immature lymphocytes into T cells.

Since these cells are essential in initiating the response of the immune system to infection, without T cells, the cascade of defensive activities either doesn't commence or does so inefficiently, spottily. Delta autoantibodies also react with the skin's Langerhans' cells, which constitute an early-warning device against infection. Both deficits are reminiscent of the genesis of much of the immune system dysfunction in AIDS patients, but this awaits a better understanding of autoimmune mechanisms secondary to infectious diseases and more knowledge about the agents involved.

There is great geographic variation in the distribution of delta infection. Pockets of delta cause considerable sickness in the Amazon basin and in central Africa. In southeast Asia, delta is confined almost exclusively to intravenous (IV) drug users in Malaysia, and the same situation prevails in the United States, where IV drug abusers are becoming infected more frequently.

Hans Popper of the Mount Sinai School of Medicine in New York, a specialist in liver disease, theorizes that "the delta viroid infection may have started, similar to AIDS and Burkitt's lymphoma, in central Africa around the headwaters of the Nile. These regions may act as 'incubators' where high temperatures and moisture may favor the transfer of the viroid from animals to man. Experimental transfer of delta hepatitis virus infection from man via chimpanzee to woodchuck and duck and possibly back from woodchuck to chimpanzee to man supports such fantasies."

Gerin thinks that "the delta studies will tell us a lot about some basic biological processes. We have to find other examples in the animal kingdom, which should tell us a lot about other diseases."

Gerin goes on to say, "We keep thinking there must be other agents in which persistent infections serve the helper function" as HBV does for delta. This might explain unusual outbreaks of severe disease. To date that hasn't happened, but Gerin and his coworkers are convinced that it will.

Meanwhile, eradication of both HBV and delta is theoretically possible—and in one fell swoop—with the anti-HBV vaccine now available. In a 1983 paper entitled "The Prevention of Liver Cancer," the World Health Organization even goes so far as to suggest that "the evidence for an association between the carrier state of hepatitis B infection and hepatocellular carcinoma [liver cell cancer] is now sufficiently strong to justify the use of a vaccine against this infection as a means of preventing this cancer."

But even in the United States and Europe, few physicians mention to patients that they should be vaccinated. A study conducted by the Centers for Disease Control showed that only a third of physicians had given the vaccine to anyone in the preceding six months. In fact, although hepatitis B cases have climbed from 200,000 to 300,000 in the past decade, only about 2.5 million Americans have received the vaccine.

Eliminating HBV, primarily by immunizing children before they become sexually active, would in effect eliminate delta as well. This would be a rare opportunity to get rid of what Gerin calls "one of the ten most prevalent cancers in the world and one of the most frequent cancers in developing countries." But although the World Health Organization has acknowledged that HBV is a major problem, funding realities are such that no plans to repeat the smallpox vaccination initiative are being considered.

14

Facts and Fancies: AIDS and Parvo

Fantasies

One of the most unsettling aspects of AIDS has been the appearance of a major disease de novo. At first, when the epidemic was identified as the "gay plague," in many quarters the attitude was "Well, they asked for it" or "It's God's punishment." Equally callous and biased rationales were trotted out when other "exotic" populations began to succumb. Haitians and drug addicts elicited little sympathy from the person on the street, and even hemophiliacs seemed a remote, special group. But when the infection began to appear in people who had received transfusions, suggesting that anyone might contract it, abstract curiosity became tinged with a sense of personal danger, which led to much blaming of various groups.

Right-wingers blamed gays; the Soviets fed disinformation to the Third World press hinting that American experiments in Africa had caused AIDS. In much of sub-Saharan Africa permission to research and publish results was withheld because reasonable—and scientifically necessary—conjecture that the human immunodeficiency virus (HIV) had its genesis in the region seemed to imply that the western world was blaming Africa for AIDS. For years many of these countries refused to acknowledge the existence of AIDS in their people largely because of its association with homosexuality, a state of being socially unacceptable on much of the African continent.

As the epidemic grew and it became apparent that there wasn't going to be a magic bullet to set it all right, certain segments of the gay community were overcome with the idea that the federal government was carrying out a kind of homosexual genocide with a virus made for that purpose in a germ-warfare laboratory. In the general panic, the fact that AIDS existed in up-country Africa and in hemophiliacs' clotting factor VIII supply and was infecting heterosexuals was explained away or ignored.

The publisher of the weekly gay *New York Native* continued to insist that AIDS is caused by African swine fever virus (ASFV). This virus, discovered in the early part of this century in Kenya and carried by healthy wild bush pigs, kills its domestic counterparts with great dispatch. ASFV has numerous clinical features resembling AIDS, such as wasting, diarrhea, and a Kaposi's sarcoma–like effect on small blood vessels, and is spread by close contact and probably through pigs eating the remains of infected pen mates.

On top of that, infection with ASFV had necessitated the slaughter of virtually the entire Haitian pig population around the time when AIDS apparently started in that country. How the virus might have jumped from East Africa to the west coast of that vast continent without East Africans contracting it until relatively late in its spread was never explained.

Among the many other facts that were ignored was that HIV, though exploding as a significant disease only recently, has been in the United States at least since 1967, found in the stored blood of a young man whose strange death led doctors to freeze his tissues. This means that a lab virus would have had to have been created at a time before much of today's knowledge of, or ability to grow, many viruses in the test tube existed.

However, a terrified readership was all the ASFV theorist needed to badger scientists and the U.S. Department of Agriculture and to warn his readers to avoid pork products. Other novel and even old pathogens have been linked by the *New York Native* to ASFV, including the putative virus that causes chronic fatigue syndrome and the microbe that causes syphilis. These pathogens have been announced as the cause of AIDS in blaring headlines, with additional charges of cover-up and conspiracy.

The fear behind those charges was generated in part by the fact that even though AIDS was killing thousands of people and obviously spreading across the world, the nation's "Number One health priority" wasn't even mentioned by President Reagan until the summer of 1986, years after the epidemic began. Perhaps as a gesture of admiration for Elizabeth Taylor, who had become involved with fund raising for AIDS after their mutual friend Rock Hudson died of the disease, Reagan spoke vaguely about testing everyone's blood. Meanwhile, hundreds of infected people held a candlelight protest vigil outside the huge tent erected for the speech beside the Potomac River.

The general anxiety about the illness was bulwarked by another nay sayer when an otherwise highly regarded scientist from California made the news time and again in the late 1980s with assertions that HIV couldn't be the cause of AIDS and that he was not afraid of being injected with the virus. Among the numerous "proofs" tallied by Peter Duesberg of Berkeley was that the long latency period ascribed to HIV was preposterous and that high levels of antibodies in the blood of those infected a priori protected them against the virus. Both biochemist Duesberg and the gay publisher were unmoved by, or uninformed about, well-known viruses whose habits parallel those of HIV.

Facts often fail to supply convincing explanations for great misfortunes. The primitive part of the mind sweeps them away in disbelief while vague notions of sorcery come churning up, looking for someone to blame. Because the disengaged President was too slippery a target, it was a befuddled Gallo who became the focus of angst, somehow considered responsible for the inability of science to work the desired magic. Virtually all criticism of the conduct of the epidemic was in one way or another aimed at the researcher, despite his being only the head of an important but small subunit of the National Cancer Institute.

If Duesberg, the *New York Native*, and the many who followed their lead had first fastened on the parvovirus family, a better case could have made for the virus and conspiracy theories. Parvovirus, although dramatically different from HIV, has caused outbreaks of disorders in animals which are similar to those of AIDS.

Parvo has the potential to become latent, to swell lymph nodes, to cause the loss of white blood cells, and to create autoimmune problems. Also, the presence of antibodies against it can signify an unfavorable outcome for the infected. All this is much like infection with HIV.

This, plus the time frame of the emergence of a novel parvo outbreak, was noticed in 1982 by Marshall Bloom of the Laboratory of Persistent Viral Diseases in Montana, who pointed out the many similarities between the new disease and the effects of parvo infection in animals. Perhaps more significant is the fact that while most people are involved with pigs only in various cooked forms, parvo infects animals with which most of us have frequent contact: cats and dogs.

Parvo de Novo

De novo is a loaded expression in research circles. *New* means that an awful lot of explaining, proving, and mind changing is going to have to be done before the new theory is legitimized and accepted. As a rule, scientists are extremely wary about making claims of or accepting anything as new. They are too aware of the "yes, buts" that greet claims of novelty as well as the number of times discoveries have turned out to be the results of manipulation of cells in the artificial environment of a laboratory or a technical glitch in a procedure. No matter how promising the apparent results of an experiment may be, if they can't be reproduced by others, they have to be regarded as a fluke.

Time and again throughout this book, quotes from researchers indicate their caution about presenting new results before "all the holes are plugged." You have often read that this or that finding has been "published," always in a peer-reviewed journal, in which any report of a new finding includes a section detailing the methods and materials involved in the experiment so that others in the same field can—or sometimes cannot—produce the same results.

The dangers of not doing so before the lay press gets its hands on a "breakthrough" came crashing down on the two Utah chemists who announced in mid-1989 that they had discovered cold fusion, which would have solved the world's energy needs. The excitement was immense. The metal palladium was an important part of the experiment, and its price soared on the stock market. But the results the two chemists had reported were deemed irreproducible by other scientists, and the excitement faded quickly. Back to the drawing board: It was nothing new.

However, the death of massive numbers of dogs in 1978 was decidedly new. Between spring and fall, a vicious dysentery tore across the United States, Canada, Australia, New Zealand, and South Africa; a year later it erupted in Europe. All types of canines were susceptible, including coyotes, maned wolves, and crab-eating foxes. Tests of frozen blood from even a year earlier showed that no canines had even had antibodies against this virus, meaning that while they might have come into contact with it before, they had not been infected with parvo.

On five continents, 80 percent of infected dogs were succumbing to a virus whose genesis could be assigned to a narrow point in time. Even more remarkable was the fact that the dog virus was almost indistinguishable from the well-known parvovirus of cats.

A similar disease had been noticed in raccoons in 1938, and a major outbreak of parvo in mink began as abruptly in the 1940s in Canada as the dog epizootic (an epidemic in animals) began thirty years later. But before 1978 none of these parvoviruses could be made to infect even the cells of dogs. What had happened?

It is "likely that the source of virus could have been cells experimentally infected with FPLV/MEV [cat/mink] and handled together with canine cells in one and the same laboratory," writes German researcher Gunter Siegl.* Siegl speculates that every veterinary product coming from the lab, vaccines in particular, in which the change might have occurred must have been tainted by the accidental cross-contamination of cat and mink viruses mingled with canine cells. Alternatively, a combination of viruses might have occurred during a deliberate manipulation of viruses and cells done to speed the production of vaccines.

In what Siegl calls yet another enigma of this panzootic, antibodies against the new virus were found in 1977 in Belgium and France even though the disease occurred in those countries a year later than it did among other affected dog populations. How could

*Gunter Siegl, *The Parvoviruses* (New York: Plenum, 1984).

French and Belgian dogs have been infected in 1977 and for two
years have shown no signs of illness while their cousins all over the
world were dying from an identical virus? But Siegl's bottom-line
concern is the unanswered question, which he calls "highly worrying": Does this tough little virus, which can live for months in dried
dog feces and which infects moth, cow, shrimp, and human cells,
have the potential to spread to other unrelated species? In fact, one
type of parvo has evidently caused illness in people for many years.

Parvo in Humans

An illustration from a book of dermatology published in 1800
shows an otherwise healthy child with what appears to be the mark
of a hearty slap on her cheek and a lacy rash running along her
arm. "Fifth disease," which is listed with other exanthems (skin
eruptions) of childhood such as measles, scarlet fever, rubella, and
roseola infantum, was described in 1889 as a kind of atypical rubella.
Though it was recognized as a separate illness thirty-five years later,
only in 1975 was a virus that would eventually be connected with
the condition found.

The pathogen was discovered by accident during tests of blood
serum for hepatitis. Not much attention was paid to it, and for
several years it was merely a virus in search of a disease. But as has
been the case with other such discoveries, once an agent was in
hand, a causal relationship between parvo and a number of conditions began to become clear. The first connection made was to
fifth disease, which was definitely linked to parvo in 1983.

Fifth disease ordinarily affects children and causes a few days
of fever, headache, malaise, rash, and sometimes the slapped appearance of the child in the illustration. Infected adults may briefly
suffer from painful joints and swollen glands. Tests of blood donors
show that 30 to 60 percent of adults have antibodies against this
virus, which is called B19. Although few people have heard of B19,
many have had fifth disease and thought it was just another late
winter flu. But far more serious connections between B19 and human health have been established. One is the serious problems the
virus causes in people who suffer from sickle cell anemia.

Sickle Cell

The genetic potential for the sickle cell affects 8 to 13 percent of black people. If a healthy man and woman who both carry the sickling trait in their genomes produce children, there's a high likelihood that the children will be born with the condition called sickle cell anemia (SCA). SCA probably confers a survival advantage in areas where malaria is endemic, because the red blood cells in which the parasites breed, when sickled, may discourage virus reproduction. However, the consequences of the aplastic anemia caused by SCA in children are often dire.

The crescent-shaped sickled red cells are fragile and able to carry less hemoglobin—which supplies iron and oxygen to tissues—than is needed. Instead of slipping through small blood vessels in an orderly way, as normal cells do, sickled red cells are irregularly shaped and tend to become stuck. This creates problems ranging from acute stomach pains to the death of unnourished tissues, particularly in the joints, and a failure of the forming bones of children to grow properly. Severe arthritislike pains accompany SCA, and neurological problems may occur because of clogged vessels in the brain. None of these problems are necessarily serious until a "crisis" occurs.

For a long time it had been suspected that the crisis of SCA had an infective cause because of the vague "viruslike" illness that precedes it. Moreover, these crises tend to occur in epidemic outbreaks at four- to six-year intervals and to affect all family members with SCA at the same time.

The first hint that there might be a connection between sickle cells and B19 emerged during the search for a way to keep the virus alive in laboratory cultures. In 1985, B19 was grown in the lab in human erythroid (immature red cells) bone marrow cells. These cells, which made up 70 percent of the total cells in the test tube at the beginning of the experiment, were reduced to only 10 percent within nine days. In other words, the virus was killing or preventing the maturation of the precursors of hemoglobin-carrying red blood cells.

The ability of B19 to "freeze" red cells and the different outcomes of this in normal people and those with SCA make it clear how important heredity can be in determining the interaction between a virus and its host. The genetic program that causes red blood cells to form the sickle shape also programs them to survive only ten to fifteen days. Ordinarily, red blood cells live around 120 days, so the virus's effects on the blood system are hardly noticed when B19 infection occurs in healthy people.

But in the short-lived red blood cells of patients with SCA, this is a critical event. Because the cells aren't replaced rapidly enough, the patient's body literally suffocates from a lack of oxygen. (Parvo causes similar problems in people with other hereditary red blood cell conditions, such as the thalassemia found in Mediterranean people.)

There is no effective treatment for any of these red cell disorders. Therefore, a vaccine against parvo—one of which has been made to protect cats and dogs—seems to be of considerable importance in preventing the destructive crises the virus causes in susceptible people, because sickle cell and fifth disease may be just the beginning of the problems parvo causes in humans.

The widespread distribution of the B19 virus has been found to complicate other diseases. Because young children often haven't been exposed to and generated antibodies against parvo, those who are treated with immune-suppressive drugs for cancers as well as babies born with inherited immune deficiency diseases can't mount an adequate antibody response and thus may become chronically infected. As a result, their already serious problems are compounded by the severe anemia created by B19 infection.

When pregnant women contract parvo, about 10 percent of their fetuses become infected when the virus crosses the placenta. B19 is also known to cause spontaneous abortion or stillbirth and may interfere with the normal development of a baby by disrupting the production of red blood cells. Hamster fetuses that survive an attack of rodent parvo often have the features of Down syndrome, such as dental defects, an infantile skull with abnormal proportions, behavioral changes, and reduced learning capabilities. Whether this also occurs in some cases of human deformity and learning disability is still speculative.

Now that we know how widespread B19 is, this has to raise questions about whether pregnant women should work in day-care centers or hospitals, where a surprising number of adults have contracted the infection from youngsters.

In 1988, health-care personnel in the Childrens' Hospital in Philadelphia were examined after two consecutive outbreaks of an illness with a rash that resembled that seen in fifth disease. The affected workers had cared for two adolescents admitted a month apart with sickle cell crisis. More than 30 percent of the caretakers who came in contact with these children became infected with what turned out to be parvo. The major symptom, the one that was most often cited as causing the health-care personnel to miss work, was swelling and pain in the joints—in other words, a transitory but severe arthritislike illness.

Although many viral infections are accompanied by joint pain and swelling, there are indications that parvo may play an initiating role in the development of one or more of the serious autoimmune conditions of which rheumatoid arthritis is the most widespread crippler. Does B19 serve as an initiating event that in some people eventuates in autoimmunity? Does some underlying genetic difference, although not as obvious as that of SCA, generate a response to this infection that sets off the self-perpetuating cascade of attacks on some part of the self that characterizes autoimmunity?

15

Immune Against the Self

In many respects the health of an individual is a balancing act between the environment and the accuracy with which that individual's immune system recognizes and deals with dangerous substances and organisms. It must be able to distinguish those that represent a threat from those that do not. At the same time, the immune system must tolerate antigens that are part of the self. The balance between the opposite ends of this spectrum—a vigorous immune attack and tolerance—is often very delicate and easily upset.

The immune system specialties with which we're endowed are predicated on our elaborate, unique sets of genetic markers. These markers clearly predispose us to or protect us from a multitude of external influences, such as our responses to stress and polluted air; the effects of smoking, food, and allergies; and our susceptibility to various infective agents.

Allergies are caused by the immune system's misreading of the potential danger of otherwise innocuous substances such as ragweed, dust, and bee venom. Those who are allergic are plagued by irrelevant immune responses, which may be as trivial as a few sneezes or can be life-threatening to the rare hypersensitive individual. In any event, it is the immune system itself, rather than the allergen, that causes symptoms. This is called a type I "immediate hypersensitivity" reaction and involves the binding of the molecules of the antigenic substance—e.g., pollen—to the surface of the special mast cells, which are found in the tissues and are already coated with a class of antibodies responsible for the symptoms of allergy.

The antigen that binds to mast cells triggers the release of histamine, which has a range of effects that cause the easily recognized symptoms of allergy. The degree of the allergic response and the tissues involved—for instance, lung tissues in asthma or those of the skin in dermatitis—are usually directly attributable to a genetic predisposition.

So is the individual response to most infective pathogens, but unlike allergies, pathogens bring into play other components of the immune system. We have seen that the initial recognition of a virus occurs only when a portion of the virus combines with a self component—the Human Leukocyte Antigen (HLA) class of molecules—to make a new "fingerprint" for presentation to immune cells. If all goes well, the T and B cells deal with the infective agent and, together with elements such as complement and various factors produced by the immune cells, eliminate the pathogen completely.

However, the balance is upset if the combination of foreign antigen and self antigen ends up resembling the self too closely. The pathogen then may be tolerated as is the normal tissue it imitates. When it goes unnoticed, a parasite is free to cause damage.

If, by contrast, the HLA signature plus the virus particle happen to resemble markers on the cells of an organ or another set of cells, the balance is tipped in the opposite direction. The self's defense then includes autoantibodies, which, as the name implies, are antibodies made against some component of the self. This may or may not develop into autoimmunity, in which there is a chronic attack on one or several tissues of the self.

Feeding into the development of autoimmunity is inheritance, which includes the HLA type, the age, and the sex of the person (in general females are six to nine times more susceptible to autoimmune diseases than are males), as well perhaps as some weakness of the target organs that predisposes them to become affected.

From another direction comes the influence of the bacteria, virus, or other pathogen that initiates a response which, as we've seen in the case of different viruses, can have a multitude of outcomes. These pathogens can cause acute or latent infections, interact with cells already infected with another virus, and so on.

Finally, there is the type and quality of the response of the immune cells themselves, which probably depends on a combination of the first two factors. Patients with multiple sclerosis, for instance, apparently suffer from fewer routine infections than does the general population, indicating that their immune systems respond in a superior way to ordinary viruses.

Though there are hundreds of conditions with an autoimmune component and though many others have been shown to be strictly autoimmune, it's hard to get a fix on what starts the process that results in disease. The bias or interests of researchers add to the confusion. Rheumatologists are eager to find a rational basis for the horrendous diseases with which they have to deal and are too seldom able to control; virologists tend to see viruses behind most obscure conditions; and immunologists, bombarded by new discoveries in their field, are inclined to be skeptical of everything.

Nevertheless, all agree that many important human diseases whose underlying causes remain unknown probably have an autoimmune component. In some, such as systemic lupus erythematosus (lupus), the attack on the self is broad and involves many different tissues and organs, including the kidneys and brain. Narrower autoimmunity is seen in insulin-dependent diabetes, in which various endocrine organs such as the thyroid and the pituitary tissues incur damage. Multiple sclerosis results from the covering of nerve cells being attacked and damaged because it is misrecognized as foreign.

In myasthenia gravis, the assault is launched against receptors on brain cells that recognize acetylcholine, a chemical necessary for the transmission of nerve impulses. Some of these disease conditions are created by specific T cells; others depend on a chronically activated inflammatory response.

The Inflammatory Response

Inflammation is the complex response of tissues to injury; by and large, it plays an important role in the healing process. The cells and the powerful chemicals they produce cause the swelling, redness, and soreness of a healing wound, a sprained ankle, or an infection. In the case of a viral infection, some of these chemicals attract or recruit phage cells to the site of infection, where the phages ingest the invading organism and regurgitate yet more toxic substances, which add to the disease symptoms.

While inflammation itself may be sufficient to deal with a bacterial infection, if the process becomes chronic, a continuous attack is mounted against the tissues of virtually any organ in the body, from the brain to the bones to the heart. This may happen when infection with a virus persists or when tissues are marked in such a way as to be misperceived as foreign.

Among the important molecules that take part in the inflammatory response are a group of twenty proteins, called complement, which circulate in an inactive form in the blood plasma. Complement is coded for by another of the HLA genes, called class III, about which not much is known.

When activated by a bacterial or viral infection or by antibodies, the complement proteins function together in an orderly cascade. Their actions increase the permeability of small blood vessels, attract and hold cells at the inflamed site, and attack the membranes of infected cells. Complement alone is known to destroy some viruses that have envelopes.

Many of the symptoms of a viral infection may be related to the activities of these complement proteins, which cause other cells to release various substances such as histamines and interleukin-1 (IL-1, or pyrogen). People born with congenital deficiencies of various complement proteins are at risk for recurrent infection and rheumatic diseases. But in fact, the way the inflammatory response is generated, along with the way in which the numerous chemical factors—of which only a few are mentioned here—are integrated, is incompletely understood. What is well known, however, is the extent of damage caused by these chemicals when they are chronically activated. This is most evident in the inflammatory conditions that fall under the general heading of rheumatic diseases, which include the western world's number one cause of pain and disability, arthritis.

Arthritis and Other Autoimmune Diseases

One of the great grab bags of medicine is designated the rheumatic diseases. British physician William Heberden (after whom the arthritis nodes that disfigure older people's finger joints are named) wrote in the late eighteenth century that "rheumatism is a common name for aches and pains, which have yet no peculiar appellation, though owing to very different causes."* The situation is little changed today. Although more than 100 discrete arthritis-associated conditions have been identified and given specific names, for the most part their causes remain unidentified.

A few are known. Diseases as diverse as gout—caused by an accumulation of uric acid crystals in a joint—and the rheumatic fever of children that once frequently followed a streptococcal infection are considered rheumatic and are included in the many forms of arthritis that affect more than 30 million Americans.

Arth comes from the Greek word for "joint," and *itis* means "inflamed." Although we tend to think of arthritis as referring to one of several specific arthritides—such as osteoarthritis and rheumatoid arthritis—many infections produce identical, if transitory, joint swelling, tenderness, and aches and pains of the type associated with classic arthritis. Syphilis and many bacterial agents can cause these symptoms, but it is viral infections that routinely mimic—and probably initiate—many cases of chronic arthritis.

So far, this book has featured viruses that cause acute infections and others that hit and hide, such as the herpesvirus and retrovirus families. The viruses that hit and run are doubtless responsible for kicking off some of the autoimmune arthritis diseases, but proving this is another matter.

The most common form of arthritis is osteoarthritis (OA), which is found to some degree in 37 percent of adults. This chronic and highly variable form of arthritis is often designated *degenerative* because affected joints are damaged and sometimes completely destroyed, particularly in older people. OA runs a gamut of outcomes from the inherited tendency mainly of women to develop the painful and unsightly Hebreden's nodes on the first joints of the fingers to severe damage of one or more weight-bearing joints. The latter may

*William Heberden, *Primer on the Rheumatic Diseases*, 9th ed. (Atlanta: The Arthritis Foundation, 1988).

result from an inherited deficiency of heat-shock protein which prevents the deep joints from handling the great heat and pressure that develop in them with use.

In many instances OA has an early inflammatory component, which is limited to the affected joints and has no effect on other organs. Viral infections may be accompanied by transitory aches and pains similar to OA, but it is unlikely that infection plays much of a role in causing the development of chronic OA.

In contrast, the classic autoimmune rheumatic diseases are characterized by frequent occurrences of specific genetic markers and by the effects they can have on many organ systems. Rheumatoid arthritis is best known for its ability to literally destroy the joints; systemic lupus erythematosus damages many organ systems, including the kidneys; and ankylosing spondylitis (AS) fuses the spinal vertebrae. All these are chronic inflammatory autoimmune diseases.

Although women have the great preponderance of autoimmune disorders, a few, such as AS and the closely related condition known as Reiter's syndrome, almost exclusively affect young men. Lupus, which strikes many more black women than white women, and rheumatoid arthritis usually go into temporary remission during pregnancy but return with grim regularity about three months after the baby is born. This connection with fluctuating levels of female hormones unfortunately hasn't led to improved therapy.

One important feature of autoimmunity is the development of antibodies against self tissues, called autoantibodies. Some researchers think, as Hugh McDevitt of Stanford University does, that "it is quite possible that humans, mammals, and vertebrates in general have been selected during the course of evolution for MCH and Ig genotypes [genetic markers] that mediate an optimal immune response to pathogens in the environment."* The rationale underlying this idea is that autoimmune conditions, particularly those involving a genetic component, are an overly successful response to pathogens in the environment by people who bear susceptibility markers. (Interestingly, schizophrenics are underrepresented among patients with rheumatoid arthritis.)

For example, there are at least forty well-described diseases in

*Hugh McDevitt, *Autoimmunity and Autoimmune Disease*, Chichester, England: 1987.

which susceptibility is influenced by the class II HLA genes, among them the skin condition psoriasis, psoriatic arthritis, lupus, chronic active hepatitis, rheumatoid arthritis, multiple sclerosis, and diabetes in children (1 child in 1,000 develops diabetes).

Abner Notkins of the National Institutes of Health reports that recent evidence suggests that a genetic component, the type of virus, and probably the development of persistent infection together play a role in many cases of juvenile diabetes (JD). Although JD appears to begin abruptly, gradual damage over several years probably precedes it. In older people, diabetes generally appears gradually and is not nearly as severe. Although their outcomes are comparable— e.g., an inability to handle glucose—adult diabetes and JD probably are different diseases.

Since their discovery a few years ago, the slow viruses are being looked at as perhaps initiating some forms of diabetes. But suspicion also falls on coxsackie B virus and persistent infection with rubella, antibodies to both of which are often elevated in diabetics. The herpes family member cytomegalovirus has also been implicated as playing a triggering role in diabetes, since it can destroy the cells of the pancreas and limit their ability to produce insulin. As is true in many diseases, damage to one organ may set up a series of dysfunctions that affect the entire organism.

In lupus, autoantibodies attack a broad range of self components, including DNA. Antibodies against red blood cells, platelets, and other cells form immune complexes (combinations of antigen and antibodies), which are deposited in joints, small blood vessels, and, of particular importance, the kidneys. Although patients with lupus suffer from a multitude of problems, kidney failure is the most frequent cause of death.

In rheumatoid arthritis, autoantibodies and a chronic inflammatory process literally eat away the joints, causing deformity and extreme pain. As noted above, inflammation attracts immune system cells with their toxic chemical defenses. These cells degrade the cartilage and the lubricating fluid that lines and cushions the joints, which may end up as merely bone against bone. The degenerative process itself continues to call into play more attacks, and eventually the affected joint is "burned out." The inflammation disappears, but the joint is left so damaged that it is beyond normal functioning.

The participation of the immune system in many of these diseases is indicated by the partial success of immunosuppressive drugs such as cortisone in turning down the immune response and thus quieting the disease process. But because their side effects are dangerous, these drugs can't be given on more than a temporary basis. The anti-inflammatory drugs—of which aspirin in huge quantities, as much as 6,000 milligrams a day (one aspirin contains 325 milligrams), is the prototype—all have unwanted effects but must be used to suppress inflammation and thus joint destruction and pain.

Dr. George Ehrlich, formerly a rheumatologist at Hanneman Hospital in Philadelphia, is convinced that he has been able to abort many potential cases of rheumatoid arthritis by aggressively treating the first sign of it in patients. His theory is that instead of waiting until the disease is established and the self-perpetuating cascade of inflammatory mechanisms is already in force, he can help patients recover from the initial attack by interrupting the disease process early with high doses of anti-inflammatory and immune-suppressive drugs. This may be impossible to prove, as patients sometimes go into permanent remission on their own.

Whither Virus?

In most cases of rheumatoid arthritis it is possible to find an autoantibody in the blood that is specific to rheumatoid factor (which itself is an indication of RA). But in seronegative rheumatoid arthritis (without the rheumatoid factor), suspicion falls on an infective agent as the precipitating cause of both transitory and chronic rheumatoid arthritis.

Rubella virus has been associated with arthritis for over half this century and is known to become latent in the central nervous system and the organs. Since the attenuated live-virus antirubella vaccine has been in use, the pathogen has been found in the synovial (joint lining) fluid of many people who receive the vaccine and subsequently experience arthritis. In most instances the joint problems resolve within weeks, but sometimes—both postvaccination and after the natural rubella infection—arthritis may persist for years as a chronic disease.

A suggestive report from Guy's Hospital in London describes six patients whose arthritis was looked at with rubella in mind. All the patients had had a recent bout of "viral illness" followed by the development of arthritis, but none had experienced the typical rash of rubella. The report's authors write that "the unusual feature about the 6 cases of rubella arthritis . . . is that none would have been suspected of suffering from rubella had not specific serological [blood] and virological methods of investigation been used." This is troubling in and of itself because of the often dire deformity caused by the virus to fetuses during the first three months of gestation.

The British researchers reported that rubella virus was re-covered from the joints of all six patients and that only the single child in the study recovered fully from arthritis. The adults contin-ued to suffer from joint disease as long as three years after the study began.

An epidemic form of arthritis is caused by the mosquito-borne Ross River virus (RRV), which belongs to the same togavirus family as rubella and which is known to consist of at least seventy viruses. RRV is confined to the south Pacific, where an epidemic of transitory arthritis was first noticed in 1928; other outbreaks have occurred frequently in Australia. In 1979 the infection appeared in Fiji, and 40,000 people became ill. RRV moved into American Samoa that same year for the first time; although no one had previously tested antibody-positive, 45 percent of the population was positive by the end of 1979.

Many viral infections cause arthritislike aches and pains, but their ability to generate chronic conditions probably depends on the ge-netic predisposition of the infected individual. This presents a prob-lem of dumbfounding complexity: With so many individual variables involved, plus the untold number of viruses that are able to induce an autoimmune response, cancer will probably be un-derstood before autoimmunity is deciphered. Every time a wide-spread new virus is identified, researchers at once begin trying to find a disease of unknown etiology with which it may be connected.

After parvovirus was found to cause fifth disease in children, because it was often accompanied by transitory arthritis, a search was begun to see what effect the infection might have in older people. At once, reports of parvo-induced arthritis began to appear. Several groups of American and British researchers discovered signs of the virus in the joints of patients with a sudden onset of inflammatory arthritis but not in anyone with OA. Only a few of the adult patients had the typical slapped-cheek rash seen in children.

Some of the studies were done with groups of patients who were experiencing an acute flare-up of arthritis; in one of these groups, recent parvo infection was found in 19 of 153 people. Other studies were begun after a known parvo infection, and small numbers of adults and children went on to develop a chronic rheumatoid arthritis–like form of arthritis. Most of them had the genetic markers that predispose a person to rheumatoid arthritis, and all those in whom the actual disease has lingered are female.

Autoimmunity and AIDS

Some of the earliest signs of human immunodeficiency virus (HIV) infection are immune system changes that bring about not deficiency but hyperactivity. They manifest in a series of conditions that are identical to those of people with various autoimmune disorders: liver damage comparable to that of lupus, inflammatory skin conditions and psoriasis, thrombocytopenia purpura (fragile blood cells), and the frankly autoimmune disease Reiter's syndrome.

In AIDS patients, hyperactivity of the antibody-producing B cells—caused by the loss of regulation of the B cells by T4 cells— results in an endless stream of antibodies to many previously experienced infective agents. This proliferation of antibodies in turn encourages the formation of autoantibodies as well as immune complexes.

Reiter's syndrome, a rare disease first described in 1916, has an affinity for the 2 percent of people with the genetic marker HLA-B27, which is also the genetic signature of 90 percent of the young men who contract AS. Both conditions are thought to follow bacterial dysentery and can result in the bones of the spine and hips becoming fused. In some AIDS patients, rapid and drastic arthritic deformities of the hands and feet and the spine are sometimes seen.

Psoriasis, which often accompanies both Reiter's syndrome and AS, is frequently experienced by HIV-positive people; in some of these patients the extent of the skin disease has been described by physicians as being "beyond anything you can imagine . . . they are ravaged by sores and lesions." The standard therapeutic approach to treating these conditions is to suppress the immune system; obviously, in AIDS patients this has already occurred. Whereas overactivity of T4 cells had been thought to cause Reiter's syndrome and AS, because AIDS destroys this cell population, researchers now believe that the T8 suppressor cell may be the culprit.

AIDS patients develop a multitude of infectious processes, any one of which may trigger autoimmune disorders as well as skin and joint problems. In disorders of this kind, the arthritis is called reactive because no organisms are found in the joint fluids.

It is clearly possible that much of the major damage done by AIDS may occur through autoimmune pathways. Citing the "incredible parallel" between AIDS and well-known autoimmune diseases, David Katz, president of the Medical Biology Institute in La Jolla, California, has proposed that both antibody and cell-mediated mechanisms theoretically could account for the multitude of dysfunctions in HIV-infected persons.

Characterizing AIDS as "dyslexic self-recognition induced by HIV infection," Katz lists several mechanisms by which autoimmunity may be generated in HIV-positive people. These mechanisms include the presence of anti-HIV antibodies that cross-react with self markers, the possibility that antibodies against HIV may block CD4 molecules (to which the virus attaches) and thus prevent the needed cell-cell communication network from functioning, and the fact that HIV carries T-cell receptors on its envelope that induce autoantibodies against self-marked (HLA) cells.

Katz suggests that if the autoimmune component of AIDS proves to be a major cause of various dysfunctions, treating this aspect aggressively and early—as Dr. George Ehrlich suggests for rheumatoid arthritis—may either abort or forestall the effects that lead to severe disease. That, however, remains to be seen.

Multiple Sclerosis

Multiple sclerosis, a debilitating neurological illness that affects 250,000 people in the United States, is an autoimmune mystery inside a conundrum. In the long and tortuous history of research into the cause of multiple sclerosis (MS) there has been much to take into account. MS isn't an isolated entity among chronic neurological diseases: Lou Gehrig's disease, or ALS (amyotrophic lateral sclerosis), and tropical spastic paraparesis (TSPP), although somewhat different from MS, share its degenerative features and peculiar clustering patterns.

MS usually appears in people between the ages of twenty and forty and can take several courses. The benign disease, as opposed to more flagrant forms, may be mild or may be diagnosed only after death from other causes. One peculiar aspect of benign MS is the lack of correlation between the extent of damage to the central nervous system's structures—which can be considerable—and the patient's ability to function. The most typical MS patient, like patients with many other autoimmune disorders, tends to have bouts of symptoms followed by periods of remission; these healthy periods gradually grow shorter as the disease becomes increasingly severe over as many as twenty years. Still other patients have a precipitous course leading to death within months, but this is relatively rare.

The discovery of an unusual number of cases of MS in Key West, Florida, highlights what has been described as the "baroque complexity" of the disease. In 1981, a young woman who had been suffering from headaches, weakness, and numbness for several years was finally diagnosed as having MS by a Miami neurologist, William Shermata. Sharing the bad news with a neighbor she knew also had MS, the woman learned that several other Key West residents were similarly afflicted. When Shermata heard that there seemed to be a surprising number of MS patients in a town with a population of just over 25,000, he ran an announcement in the local paper to see if there might be more.

The result was astonishing: In all, thirty-seven local people were suffering from MS. Even this small number represented forty times the number the doctor would have expected to find with the illness in this population. Among these people, 24 percent were nurses. Equally peculiar was the appearance in the tropics of such a cluster: People born in and spending their lives in northern latitudes have a far higher risk of contracting MS than do those from the south. Even people who move from north to south after adolescence carry the risk with them, indicating that infection before the teen years must have a major role in the later development of MS.

But the great majority of Key West patients were natives; two pairs were siblings, and two others with MS had affected siblings living elsewhere. Although many of the island's natives are of mixed African-American heritage, except for a black sailor from the local navy base, all of those with MS were white.

Many questions were raised as the unusual cluster became known, largely the same ones that have been asked ever since MS was identified as a chronic degenerative condition over a century ago. Why did it take years to diagnose MS in the patient mentioned above? As is often true, in the absence of physical abnormalities to which to assign various symptoms such as dizziness, numbness, and exhaustion, patients are often dubbed psychoneurotic. Consequently, not until more obvious signs of motor dysfunction appear do most patients finally go to a neurologist.

But why were there so many cases in Key West? Suspicion first fell on the island's poor sewage facilities and open garbage dump as perhaps harboring causative pathogens or toxic wastes. Or could the heat and humidity have set off the outbreak in the genetically susceptible? Because there was a high incidence of AIDS in the island's many gay men, possible connections between the two diseases were also discussed. But none of these factors seemed likely, and experts were able to point to similar miniepidemics of MS over the years.

For instance, the disease was unknown in the Faroe Islands northwest of Great Britain until British troops arrived in the 1940s; within a few years, twenty-four cases had surfaced. Cases of MS in Iceland had also increased dramatically after European soldiers were stationed there during World War II. The possibility that an infectious agent is involved in MS is strongly suggested by the rise in the incidence of this disease in the Scandinavian countries of Finland and Norway, where between the early 1960s and the present the incidence of MS has been growing, with cases in Norway increasing threefold. Currently, the number of Israelis with MS also seems to be on the increase.

All these clusters and the growing incidence of MS in many parts of the world suggest an infectious agent at work. Because of the long latency period between preadolescent infection and disease, the agent would have to be some type of virus, which also would have to be widespread in the environment. Another requirement seems to be a genetic predisposition.

This idea has been reinforced recently by research involving forty families in which two or more members had the disease. The children who later went on to develop MS all had a particular arrangement of five genes; their healthy siblings had different genetic patterns. These "susceptibility" genes participate in the immune recognition system that enables T cells to identify targets, suggesting that a genetic inability to do so allows an unknown virus to become persistent and gradually scar myelin nerve sheaths.

Given the difficulties of diagnosis even by experts, it's likely that there is more MS everywhere than current figures indicate. When the already high incidence of cases in the Orkney Islands was more closely studied, MS turned out to be ten times more common than was thought. Diagnosis is based on symptoms, which in part depend on where the nervous system is damaged: Weakness and prickling point to spinal cord lesions, while problems with vision can be traced to damage of the optic nerve.

Many viruses have been investigated as possible causes of MS, among them the measles virus. Antibodies against measles and the measles virus itself are found in the spinal fluid and brains of about 60 percent of people with MS, but the same holds true for many normal people, so the question remains.

However, attention is being focused on the slow viruses and their possible role in many central nervous system (CNS) and "mental" diseases. The retrovirus HTLV-1, which causes adult T-cell leukemia, or a virus very closely related to HTLV-1 is beginning to be linked to MS (Chapter 17). Although there is no direct proof that this slow virus actually causes the disease, in a study of MS patients in Florida and Sweden, as many as 60 percent were found to have antibodies against this relatively rare virus, while healthy controls showed no evidence of a similar encounter.

In MS with a benign course, inapparent or negligible changes may be present for years or for the rest of the patient's life without a definitive diagnosis being made. Are people with cognitive problems characterized as mental disorders that remain undiagnosed actually suffering from MS? This may turn out to be a rational suspicion, because many viruses both subtly and obviously affect the brain with a frightening array of insults.

While the newly discovered class of slow viruses is being linked to mysterious disease entities, many well-known viruses carried by insects continue to cause diseases that we in the West seldom consider. In tropical climates, however, these diseases are common and many are deadly.

16

Animal and Human

Arboviruses

In an inspection of airplanes landing at the airport in Nairobi, Kenya, researchers found thirteen species of mosquitoes; among them was the strictly American *Aedes taeniorhynchus*, found aboard a flight that had originated in Rome. What might these international insects have been transporting? O'nyong-nyong, Chunkungunya, dengue fever, yellow fever, or another of the more than 250 identified arboviruses, at least 80 of which cause disease in humans.

The mere presence of the mosquitoes was no guarantee that they were carrying a novel virus or, even if they were, that local insects would be able to adapt to it. But the ability of these insects to serve as the vectors of disease-causing viruses has changed the history of the world.

Vector is from the Latin for "carrier" and is ordinarily used to refer to insects that transport diseases between plants or animals. Humans can also act as vectors, and the infections with which the western world's cities are most familiar are those that are passed from human to human. Given a sufficiently large pool of people—several thousand is probably the minimum—and a virus passed easily between individuals, many waves of infections will come and go.

Measles, smallpox, whooping cough, and mumps are typical infections that are passed by sneezing and breathing. People are also vectors for sexually transmitted viruses, such as papilloma, genital herpes, human immunodeficiency virus (HIV), and hepatitis.

If, however, several vectors are needed or if other animals serve as reservoirs of virus, an entirely different set of circumstances is required for a virus to gain a foothold in the population. Everything to do with viruses becomes enormously complicated when an insect or animal is an integral part of the chain of transmission and people just happen to be accidental bystanders, unnecessary to the pathogen's life-cycle needs.

Arboviruses (*ar*thropod *bo*rne) cause a range of human diseases, many of which cause changes in the brain or small hemorrhages of blood vessels. In the main, infection by an arbovirus is the result of man's inadvertent entry into the complex ecological cycle between arthropods and natural animal reservoirs. Birds and mammals are the natural hosts for the biting and sucking insects that carry these viruses, and a human being becomes an unlucky meal when wandering in their territory. Occasionally these animal viruses spill over into inhabited areas when drought drives infected mosquitoes into populated areas to breed in standing water. Mosquito-carried St. Louis encephalitis, California encephalitis, and eastern and western equine viruses most frequently cause disease in this country.

Because they depend on the territory covered by their hosts, arbovirus outbreaks are limited to specific areas. Japanese encephalitis (JE) is confined to, though widespread in, the Orient. First recognized as causing a disease in 1871, the agent was isolated in 1934 and is the only one against which a vaccine has been made. In recent years, mortality has dropped from 40 to 50 percent to 10 percent, perhaps because a less virulent strain of the virus is now more common. While pigs in endemic areas are 100 percent positive for antibodies against JE, they remain healthy, as do dogs and other animals that have been infected; only horses, donkeys, and people become ill from JE.

In his fascinating book *Plagues and People*, historian William McNeill gives an example of what can happen when a viral disease depends on an insect vector for its survival. The carrier of the yellow fever virus, the *Aedes aegypti* mosquito, was brought to the Caribbean in water casks early in the history of the African slave trade. But the region remained free of the killer disease until *A. aegypti* was able to establish a foothold several years later in the New World.

Monkeys were originally thought to be the reservoir of yellow fever in Africa, where they are unaffected by the virus. However, in the Americas they apparently had never before been exposed to the virus, as they too were killed by "yellow jack." It turns out that monkeys, like people, are highly susceptible to the yellow fever virus, which carries out its life cycle not in the primates but in the mosquitoes themselves.

The presence in Central America of both yellow fever and malaria turned the original plans for a Panama Canal into a disaster by killing or debilitating the workers. When Walter Reed finally proved in 1900 that mosquitoes were the carriers, yellow fever was controlled by eliminating the insects' natural habitat and work on the canal was finished. However, the victory over *Aedes* was temporary because a pool of infected monkeys and mosquitoes keeps the virus available.

The same pattern is being repeated in a limited way as spin-off of the cocaine trade: Bolivian Indians are leaving their mountain homes to grow coca at the edge of the jungle. As they move into the range of the carrier mosquitoes, these squatters are dying from their first exposure to the yellow fever virus.

In areas where yellow fever is endemic, most people are sufficiently resistant through early exposure so that serious illness seldom occurs. But when introduced into virgin populations, the virus causes deadly epidemics, with mortality rates of 20 to 50 percent, particularly in children. In the 1960s an unusual outbreak of yellow fever in the Sudan and Ethiopia infected 200,000 people and killed an estimated 30,000.

Viruses that are transmitted by insects are classified broadly as arboviruses. In some cases the biting insect acts as a mechanical conduit, a "flying pin" that merely transports viruses between one creature and another and plays no role in its life cycle, as was the case with the myxovirus that killed millions of rabbits in Australia. The ability of different insect species to adapt to new viruses, in fact, was responsible for the unexpected spread of the rabbit killer. Initial transmission was by a type of mosquito that haunts river courses; the following year, the virus had leapt the great stretches of desert by way of mosquitos looking for shelter in cool, humid rabbit burrows.

Alternatively, an insect may be the primary host in which a virus lives and replicates. Such viruses are transmitted through the insect's infected salivary parts. In African swine fever, for instance, infected warthogs are simply used as a supply of blood for the female tick, which passes the virus to her progeny. The infected baby ticks attach themselves to and infect a hog, which then serves as the vector as it comes into contact with other pigs. Because of their dependence on the reservoir animals, which in tick terms are the center of the universe, these pathogens are usually limited to specific ecological regions.

Hemorrhagic Fevers

Generally speaking, arboviruses can be divided into those that cause encephalopathies (inflammations of the brain) and those that cause hemorrhagic disorders. Both yellow fever and dengue fever, plus a multitude of others, are hemorrhagic and result in the jaundice and liver damage often experienced by patients with hepatitis. Rodents are the natural hosts of many of these viruses, and all of the ones that infect people have the potential to be dangerous.

In China during 1980 and 1981, there were 30,000 and 42,000 cases of hemorrhagic fever respectively, requiring hospitalization and carrying with them fatality rates up to 7 to 15 percent. Eleven thousand cases occurred between 1980 and 1985 in the far eastern part of Russia, and deaths still occur on our doorstep in Cuba, Nicaragua, and Aruba.

Dengue fever has been recognized as a clinical entity for at least two centuries, as epidemics recorded in Java in 1779 and in Australia in 1897 show. The disease is a first cousin of yellow fever and shares the same mosquito vector; both viruses belong to the arbovirus group that causes hemorrhagic disease. While other hemorrhagic fever viruses, such as Lassa fever, Marburg, and Ebola, have received much public attention because of their novelty, the widespread old killer viruses are the ones to watch.

In the northern hemisphere we give little thought these days to these animal-insect diseases, but our memory is short. Five hundred thousand cases of dengue fever were diagnosed in Texas in 1922. With the recent migration of the aggressive Asian tiger mosquito into Brazil, we may be facing a resurgence of this infection. Eleven thousand cases occurred in Puerto Rico in 1986, and there were 89,000 cases in Latin America in the same year.

The most significant difference between yellow and dengue fevers is that natural exposure to any variety of yellow fever confers immunity, and vaccination protects one against it for about ten years. But dengue fever is another matter. There are at least four distinct serotypes, and infection with one strain—while it protects against that strain—leaves the individual open to becoming ill with another.

An initial dengue infection causes high fever and a rash, headache, eye pain, and arthritislike aches of muscles and joints that are incapacitating (dengue is sometimes called breakbone fever). Although it may take a long time to recover fully from dengue, the first infection is seldom fatal.

Hemorrhagic dengue fever is often a fatal disease and is the leading cause of hospital admission of children in southeast Asia. Some think it may be caused by a more virulent viral strain, but it seems likely that the immune status of the host is critical and that antibodies developed during previous dengue infection or antibodies passed from mother to infant are the real culprit. The theory is that antibodies against the earlier infection or those passed from mother to infant interact with the next virus, perhaps by making it easier for it to penetrate macrophages and monocytes. These vital cells, now decorated with the viral antigens, become the target of the immune system attack. When this occurs, shock and hemorrhage follow.

During the years I lived in Kenya, for political reasons it was impossible to cross our contiguous borders with Uganda, Tanzania, Ethiopia, and Somalia, but it was the green monkey virus that kept the Sudan off limits. The deadly sickness had appeared de novo only a few years before just across the border from Kenya and, at the same time, 850 kilometers away in the high forests of Zaire. The virus has been named Ebola. In late 1989 and early 1990, Ebola virus was found in a number of monkeys imported by primate research facilities. This has necessitated new guidelines to protect lab workers from the deadly pathogen.

Ten years earlier, a devastating outbreak of hemorrhagic fever among laboratory workers in Yugoslavia and Germany was quickly traced to infected tissues of monkeys flown in from Uganda. Seven of the thirty-one workers died, infected with a virus now called Marburg.

But are these really new diseases? In the years since their appearance, investigators have found that the Ebola and Marburg viruses are closely related and that various African populations carry antibodies against them: People in Kenya, Uganda, and Zimbabwe have evidence of having been infected with Marburg. In Cameroon on Africa's west coast, almost 10 percent of young adults are antibody-positive for Ebola, for which up to 33 percent of rain forest pygmies also test positive.

The viruses have been looked at, characterized, and given a new classification—filoviridae, for the bizarre, threadlike appearance unlike that of any other animal agent. But despite the intense interest of researchers in these abrupt, deadly outbreaks, no animal reservoir has been firmly identified. Ebola kills monkeys, which rules them out, so it is unlikely that it or Marburg harmlessly "winters over" in primates.

In both diseases, hemorrhages and liver destruction lead to death within two weeks, but in Africa as in other developing areas in the southern hemisphere, hemorrhagic fevers are an unfortunately routine mattter. The animal viruses that cause hemorrhagic disease in humans are found everywhere. So, unfortunately, are those that affect the brain.

Encephalitis

Viruses gain access to the brain by several routes. One is direct inoculation into the blood, as happens when they are transmitted by animal bites (rabies) or insect-carried pathogens. HIV is thought to gain entry hidden in the occasional macrophage that ends up in the brain. Other routes include the respiratory tract, where nerve fibers in the nose provide a conduit between the external environment and the brain.

Other traveling arrangements include crossing the placenta from mother to fetus, as can happen with rubella, parvo, and cytomegalovirus infections. These infections are termed congenital, and a high percentage of them result in deformity, mental retardation, and other types of neurological impairment.

The usually deadly rabies virus, to which all warm-blooded animals are susceptible, has the most dramatic outcome of any acute viral infection of the central nervous system (CNS). This RNA virus, which can sometimes be contracted by inhalation as well as from the bite of an infected animal, may incubate for as long as several months before symptoms appear. But at some point it begins replicating in muscle tissue and gradually reaches exposed nerve endings.

Moving along the nerves, the virus ascends to the brain and, after growing in certain neurons, travels back down the nerve cells. Once this process is under way, the classic frenzied rabid behavior, almost always resulting in death, follows quickly. Because so many wild creatures carry rabies, there is no possibility of "le rage," as it was once called, being eliminated.

Many acute viral infections as well as other agents and toxins are known to cause hallucinations, loss of motor control, coma, and death within days to weeks. Because of the implications of invading the brain to look for these agents along with the difficulty of actually finding the virus at autopsy, the causative agent is seldom identified.

Encephalitis, or inflammation of the brain and the CNS, is the most frequent outcome of acute brain infection and involves the meninges, the brain's covering tissue. Although this serious disorder can be caused by bacteria and fungi—in fact, by many of the same opportunistic pathogens that infect patients with AIDS—viruses, especially arboviruses, are responsible for the majority of encephopathies.

In most infections that affect the CNS, the mortality rate is about 5 to 10 percent, but in some herpes infections and in eastern equine encephalitis, mortality can be as high as 70 to 80 percent. Herpes simplex is the only one of these infections for which treatment is available.

Encephalitis can also follow vaccinations against a number of viral agents, including vaccinia, yellow fever, rabies, and, as mentioned earlier, polio. These rare events are thought to be brought about by an immune system attack on cells of the CNS, which is true of a number of CNS conditions. Even in rabies, because the actual damage to the CNS is inadequate to account for the lethal outcome, Dr. Hilary Koprowski of the Wistar Institute in Philadelphia speculates that suppression of T cells generated by the virus itself plays a role in the clinical symptoms.

Alternatively, or in addition, antibodies that attack myelin, the fatty covering of nerve axons in the brain and CNS, may account for some of the effects of rabies. When rabies vaccine is made in animal brain tissue, 1 person in 500 receiving it experiences neuroparalytic accidents (as opposed to natural infections). In these cases, the antibodies made against the basic protein of myelin show that the brain tissue used to manufacture the vaccine causes the recipient to mount an immunologic defense response against his or her own brain tissues.

Because human CNS nerves are also covered by myelin, this essential substance is misread as foreign and is eroded by the autoimmune attack. When the vaccine is made in duck embryos, which contain no neural tissue, neuroparalytic accidents are absent.

The symptoms of postvaccine accidents are identical to those that follow natural infection with the measles virus: One patient in a thousand develops antibodies and T-cell responses against myelin nerve sheaths. In most cases, these are temporary conditions, but when they persist, chronic CNS disease follows.

The length of time an acute viral encephalitis lasts and its outcome depend on the agent involved, the severity of the illness (whether there is coma, seizure, etc.), and the host's individual immune response. Typically, there is a long recovery period, which, in the case of California encephalitis and western equine encephalitis, usually resolves without residual problems. However, other viruses aren't so forgiving; infections with eastern equine virus and herpesvirus often leave high mortality and morbidity in their wake.

All this sounds cut and dried, as if infection were followed by certain symptoms, diagnosis, and one of several outcomes, but it is far from being so. Although many of the viruses that cause mental disturbances are well known, suspicion is mounting that others may be responsible for what we are used to thinking of as psychiatric problems. Many researchers are trying to identify unknown agents that may be floating around the world causing psychiatric problems. One of these is the Borna disease virus, which is known to cause behavioral disorders in animals.

Borna Disease Virus: Cause for Alarm?

In 1894, an outbreak of deadly encephalitis in horses and sheep occurred in the German town of Borna. Although this animal disease has never been found outside a few small German and Swiss rural areas, the effects of the unclassified virus (BDV for Borna disease virus) on experimental animals, as well as its predilection for the limbic area of the brain, where depression is known to originate, have led researchers to look at it as a possible cause of human depression.

The question was put this way: Since in animals BDV induces first frenzied, socially disruptive behavior and then an apathetic phase—equivalent to manic-depressive disorder in people—might it cause similar symptoms in humans? Without having been able to characterize BDV, researchers nonetheless knew that it had the qualities of other slow viruses. BDV doesn't kill cells, is spread by cell-cell contact, and has a long incubation period; antibodies against it fail to control infection.

Rats infected with BDV show no clinical signs for about three weeks; then they become what German researchers describe as "alert."

"During the next few days alertness progressed to a state of frenzy. . . . They seemed disoriented and their running and jumping lacked coordination."*

Gradually, the rats became aggressive and attacked their cage mates. The animals ate ravenously, and the males had constant erections (a not uncommon condition in some human mental disorders).

The researchers reported that this behavior gradually receded, after which the animals "became passive and showed gradually less activities during the ensuing weeks."

In experiments, many animals, including primates and birds, can be infected with the virus. In rabbits it causes a fatal encephalitis, and in some monkeys BDV causes a chronic encephalitis with social and behavioral abnormalities.

Although BDV has never been known to cause human disease, because of the similarities of its effects to human manic-depressive disorders, researchers in Pennsylvania decided to look at the blood of depressed patients to see if any of them had signs of BDV infection. The first search was done in patients with major depressive

*E. Kurstak, Z. Lipowski, and P. Morozov, eds., *Virus, Immunity, and Mental Disorders* (New York: Plenum, 1987).

disorders in the University of Pennsylvania Hospital, comparing them with healthy people from the community.

The results were that 4.5 percent of the 265 patients, but none of the controls, had antibodies against BDV. In later studies, 2 percent of patients had high levels of anti-Borna antibodies, while the same percentage of controls had low levels. Interestingly, 8 percent of those positive for HIV infection also had antibodies against Borna.

The researchers raise the caveat that perhaps the altered immune system response of both HIV-positive and depressed people predisposes them to infection with a variety of viral agents rather than the other way around. This argument is easily raised about viruses to which many people in the same area are exposed. But the presence in depressed patients of antibodies against a rare neurotropic virus tips the scale of probability toward BDV playing a role in some cases of depression.

Many questions have to be answered before a genuine association between BDV and human mental conditions can be made. The first that comes to mind is, Where did people who had never left the country contract Borna? Instead of Borna, perhaps what the researchers have been measuring is the presence of a virus closely related to it, which is cross-reacting in the antibody test. Of course, even if it is Borna or a close relative, how can we prove that it causes depression? Ideally, by more or less the same kind of prospective studies used by the Henles when they showed that Epstein-Barr virus (EBV) causes mononucleosis: by studying a large, uninfected population and seeing who gets sick and noting whether the virus appears at the same time.

But the Henles knew that the virus they were studying was everywhere and easy to catch through casual contact, while BDV is so rare that few people would ever come in contact with another person infected with it. There are also significant differences between identifying a virus that causes a specific kind of illness with distinct symptoms, such as the fever and swollen lymph nodes of mononucleosis, and the relatively vague feelings of depression virtually everyone experiences.

Too little is known to draw any firm conclusions about BDV as a cause of some mental disorders, but many people with what initially appeared to be depression, schizophrenia, or psychosis have been hospitalized. Only later, when psychoactive medications have failed to moderate the emotional component and recognizable viral-caused disturbances in consciousness have begun to surface, have these patients been diagnosed as suffering from organic rather than mental conditions. Called "the great imitator," virus-caused encephalopathies often mimic psychiatric symptoms, with hallucinations and delusional ideas appearing long before any signs of physical illness become apparent.

This may raise a diagnostic dilemma that is, or perhaps should be, posed when formerly balanced people suddenly begin to experience mental disorders. The brain, in its expression through organized, coherent actions and thought processes, is far more open to viral interference than we once thought.

PART FOUR

VIRUS AND THE MIND-BRAIN

17

Virus and the Brain

The seventeenth century's collection of intellectual and scientific innovators included the philosopher René Descartes, who concluded that a person is merely "an earthen machine," whose only proof of existence depends on the ability to think. Descartes summed up his philosophy with the phrase *cogito ergo sum*, "I think, therefore I am."

Those who are trying to figure out how the thinking organ works, however, are less sanguine about what they understand.

"What we know of the brain is like dropping a microphone down into a single room in a ten-storey office building and trying to interpret what's happening in the whole building by one overheard conversation," sighs Dr. William Slikker.

Slikker and his colleagues at the National Center for Toxicological Research in Little Rock, Arkansas, are attempting to define what causes the aging process.

"The complexity of this is enough to drive a lot of people from the field," says Slikker, who nonetheless maintains that "the loss of the major homeostatic organ—the brain—seems to be the key to the process of aging."

"We're really just cracking open the box to look in the brain and trying to understand what all of these receptors for most hormones are doing in there!" he explains.

Today's philosophers continue to wrestle with the problem of having to use the brain to think about the brain and how it produces creativity or intelligence. But it is the anatomists, physiologists, chemists, neurologists, and psychiatrists who now provide insights into the brain's intricate electrochemistry and the central nervous system that is responsible for relaying its messages.

As reductionist as Descartes's statement is, it nonetheless expresses the criteria for an intact individual. If "I am" because "I think," when the unique individual is gone from the "earthen machine" in those who are declared brain-dead, the "I" no longer exists. We are what we are courtesy of the unique wiring of the collection of soft, convoluted tissue with its electrochemical network safely tucked away behind the blood-brain barrier.

The Blood-Brain Barrier

Since early in the nineteenth century it has been known that some kind of barrier separates the brain from the body's general circulation. Renowned bacteriologist Paul Ehrlich injected dye into the veins of animals and observed that although it circulated freely into all the other organs, their brains remained untouched. Ehrlich interpreted this as showing that the dye had no affinity for brain cells.

Years later, Ehrlich's former student Edwin Goldmann reversed the experiment and injected dye into animals' spinal fluid. This time the animals' brains but not their blood changed color. The combination of the two results proved that a barrier exists, but Goldmann drew the erroneous conclusion that the barrier is impermeable.

Gradually, it has been learned that a membrane made of capillaries separates the brain (and its branching central nervous system) from the circulating blood. Unique among body tissues, the capillaries of the blood-brain barrier make up a continuous wall across which few substances can pass. This makes the brain a privileged site free from most of the chemicals that affect other parts of the body.

Since we have many of the same receptors on brain cells as on cells of other organs, we are jerked this way and that by sex hormones, adrenaline, and a multitude of mood-modulating chemicals.

But if the brain were exposed to the fluctuating influences of substances such as the salts and acids that ebb and flow in the bloodstream, our mood swings and behavior would be uncontrollable.

However, the barrier can't be completely impermeable: Essential nutrients must be able to get in, and surplus substances have to be pumped out. Generally speaking, molecules that dissolve in water are prohibited, while molecules with a lipid or fat makeup are allowed to pass. The fatty molecules of ethanol, alcohol, nicotine, marijuana, and heroin, for example, are lipid-soluble and thus readily cross the barrier. However, nonlipid but essential brain nutrients, such as glucose and amino acids, are recognized at the barrier and are carried into the brain by molecules whose specific function is to ferry such substances across.

Other barrier cells have a special metabolism that changes the chemistry of certain substances so that they can enter the brain. An example of this is the way these cells handle heroin and morphine, the main products of the opium poppy. Since heroin is derived from morphine and reverts to it in the brain, it would make sense for the original substance to be used by addicts. But the process that turns morphine into heroin provides it with a lipid "ticket," allowing easy penetration through the barrier, whereas morphine is less well absorbed. When stripped of these lipid molecules at the barrier, heroin turns back into morphine.

This same mechanism works against effective therapy for patients with Parkinson's disease, whose brains cannot produce the neurotransmitter dopamine, the lack of which causes the altered central nervous system (CNS) function characteristic of the disease. Attempts to restore dopamine with L-dopa, a drug composed of the amino acids from which dopamine is formed, theoretically should be able to replace the essential molecules. But the barrier alters L-dopa in such a way that it reaches the brain changed. While the drug sometimes brings about a "miraculous" reversal of symptoms, its effects are capricious and often transitory.

Although the blood-brain barrier also prevents most medications from reaching the brain, its protective mechanisms are necessary to keep out the forest of flora and fauna that normally inhabit the body. When this barrier is breached—more often than once was thought possible—there's hell of a special kind to pay.

The most dramatic outcomes usually occur at the two ends of

the age spectrum: during infant development and as aging takes its toll on many formerly orderly metabolic processes. But we are subject to changes in brain function, which often mimic emotional and cognitive problems, at any age if the integrity of the blood-brain barrier is breached by a viral infection.

Disturbed or Diseased?

When the early cases of AIDS began appearing, the disoriented and disorderly way in which many patients behaved was assumed to be caused by their being depressed. After all, they had a dreadful collection of symptoms, were literally wasting away to nothing, were covered with Kaposi's sarcoma splotches, and so on.

Then direct infection of the brain by various opportunistic parasites was identified at autopsy. It happened this way: George Hensley, chief pathologist at Jackson Memorial Hospital in Miami, kept puzzling over the frozen sections of brain from a young man whose death couldn't be explained by his doctors. The particularly vicious tuberculosis (TB) that had killed the young Haitian "just didn't look right," Hensley recalls. This wasn't the first such slide Hensley and his people had struggled over in the spring of 1981, soon after the straggling boatloads of Haitians began to appear in our southernmost metropolis.

As was his routine with unidentifiable specimens, Hensley froze and filed the slides. These were soon joined by other equally puzzling sections as other Haitian patients died of TB with neurological manifestations such as seizures, disorientation, and confusion.

Finally, there was a section in which the distinctive beadlike shapes of the parasite *Toxoplasma gondii* slid under the microscope's eye. "Bang!" says Hensley. "We had an epidemic."

Soon the brains of AIDS patients were also found to be populated by *Mycobacterium avium-intracelluare*, which, prior to the new illness, had been reported only fourteen times in adults worldwide. Several other patients' brains were found to be infected with *Cryptosporidium*, which is so rare in humans that only six cases were known to have occurred between 1976 and 1981. But in 1981 and 1982, the late Pearl Ma of St. Vincent's Hospital in New York diagnosed it in the brains of fourteen men with the new syndrome (Chapter

2) Donald Armstrong, head of infectious disease at Sloan-Kettering Cancer Center, said, "I've been here for twenty years, and I've never seen anything like this before."

As the syndrome began to come into better focus, and particularly after the likelihood was raised in 1985 that a viral agent is responsible for AIDS, physicians began to suspect that many of the wide variety of mental and CNS changes were being caused by a virus. Numerous researchers reported finding traces of human immunodeficiency virus (HIV) in brain cells. In Boston, after other possible causes of meningitis had been ruled out, the brains of several patients were finally found to be infected with the virus. Scientists at the National Institutes of Health next showed that HIV was actively growing in the brain and CNS of AIDS patients.

The list of possible organic brain disorders in HIV-positive people encompasses the whole range of what we usually think of as mental and CNS disorders: marked changes in personality and behavior, inability to concentrate and remember, incoherent speech, uncoordinated movements, delusions, anxiety, insomnia, seizures, and delirium. Infected children fail to develop the ordinary skills or attain the intellectual milestones that should accompany growth.

By 1989 most clinicians agreed that many HIV-positive individuals, even before they begin to show symptoms of illness, may have signs of mental problems, in particular, language disturbances and memory loss. Thus one more confounding element has been added to the development of effective therapies: Drugs able to protect blood cells against HIV must also be able to reach infected brain cells.

Mind, Brain, and Central Nervous System

The differences between emotional and psychiatric, or, "mind," disorders and those caused by organic disturbances such as those that follow infections are difficult to nail down because the symptoms are often identical. For instance, in AIDS patients, although psychotic behavior, severe anxiety, forgetfulness, hallucinations, and depression mimic a number of mental conditions, the great majority of these symptoms can be attributed directly to infection with HIV.

A distinction has to be made between *biopsychosocial* and *biophysical* disturbances. The first category covers the emotional and thought disorders that are experienced by people whose problems are related to various types of personality disorders, some of which occur in response to external situations. The second category assigns physical causes to behaviors that may share similar features but are caused by structural changes in the brain or its peripheral nervous system.

For example, AIDS patients may be laboring under the biopsychosocial problems caused by society's fear and loathing of those with the illness while at the same time dealing with the biophysical insult that affects their thinking functions. AIDS patients also often suffer from ataxia, slurred speech, and jerky movements, which are neurological disorders that indicate that the CNS is affected as it is in diseases such as multiple sclerosis and Lou Gehrig's disease. Both mental and neurological functions can be disturbed by events taking place within the brain, while physical functioning is disrupted by changes in the CNS.

Over history, behavioral disturbances have been blamed on everything from possession by demons, to smelling basil (an activity once thought to cause a worm to grow in the brain), to bad mothering, to social pressures. Today, many brain infections caused by viruses are known to produce thought and perceptual disturbances.

This raises a host of other considerations. What of conditions such as schizophrenia, depression, and mania? And what of the degenerative diseases associated with, but by no means limited to, the aging process? (Some of these are considered in Chapter 20 in a category different from what we think of as viruses.)

The unpleasant probability is that many mental dysfunctions are caused by changes in the actual milieu of the brain. This is bringing the practice of psychiatry into close contact with virology and immunology, an often uncomfortable relationship but one with important implications for treatment.

For example, if the underlying cause of disordered thinking and behavior is a change in the brain, reversing or moderating this change with appropriate medications would be more realistic than searching emotional histories to explain what are currently considered to be psychiatric problems.

Psychoactive drugs such as lithium effectively smooth out the

up-and-down emotional swings of manic-depressives; major tranquilizers such as haloperidol (Haldol) can sometimes suppress the
erratic thought processes of schizophrenics; and tranquilizers and
antidepressants can reestablish a more normal emotional balance in
less troubled people. Better knowledge of deficits in, and replacement with, other chemicals may hold great promise for treating
many mental disorders.

Depending on what types of cells are damaged, viruses can have
many different effects on the brain and CNS. Some infections result
in transitory organic syndromes such as dementia and delirium;
others may not produce noticeable symptoms but instead have a
subtle impact on behavior; some may play a role in functional psychoses such as Alzheimer's disease and schizophrenia (though this
is far from proven); still others, by changing mental or physical
functions, may precipitate reactive psychiatric disorders.

These highly variable outcomes depend on many factors: the
kind of virus and the cells for which it is programmed, whether the
virus causes acute or persistent infection, and the response of the
host, which, as we have already seen, is related to a multitude of
different factors, including genetics and the age of the host when
infection occurs.

The Young Brain: Defective or Infected?

The Centers for Disease Control (CDC) recently came out with
a peculiar report which purported to explain the United States's
shameful rank (twentieth in the world) in infant mortality. As quoted
in *The New York Times*, one of the study's authors concluded, "When
we talk about changing infant mortality rates, we're not talking about
a simple intervention program."

The reason given for the agency's pessimism is that despite the
disproportionately high death rate of infants born to poor
women—mostly attributable to a lack of prenatal care—birth defects
account for 20.5 percent of infant deaths. Two-thirds of these defects are described as being of unknown origin.

A list of the major causes of death of infants younger than one
year old includes low birth weight, prematurity, and respiratory
distress, which together account for 17.7 percent. Sudden infant

death syndrome is responsible for 13.6 percent, and infections, including influenza and pneumonia and sexually transmitted diseases,
account for 4.1 percent. Combined with birth defects, these causes
together account for 55.2 percent of the almost 40,000 yearly infant
deaths in the United States.

But viruses are likely to be the "unknown origin" of many of
the birth defects. They may also play a role in the 3.6 per 1,000
babies who are born retarded and the approximately 5 million
American children with learning disorders.

Given that parvovirus B19 has been widespread in the population for a long time but only recently has been found to cause sickle
cell crisis, spontaneous abortions, and perhaps birth defects, might
not other viruses—aside from those already known to affect fetuses
and infants—be responsible for many of the defects of unknown
origin cited in the CDC study?

The precarious timing of fetal development has been highlighted by the deficits seen in infants born to mothers who abuse
alcohol during pregnancy, a major cause of mental retardation in
the western world. Babies with fetal alcohol syndrome even share
a peculiar facial structure, showing the profound effects of alcohol
on the precisely timed and orderly pattern of development. Researchers have found that animals exposed to the equivalent of
twelve cans of beer a day have offspring with mental retardation
and motor dysfunction.

The extent of injury to a fetus or newborn depends on many
factors, including the type of agent and the baby's stage of development. Obviously, infection that interferes with the formation of
organs during the first trimester of pregnancy has a considerable
potential to disrupt the orderly progression of development. In
some cases, the second three-month period is also critical, as is
infection by some of the viruses that are acquired during delivery,
such as the herpesvirus cytomegalovirus (CMV).

These examples raise questions about the disastrous condition
called autism, in which children are withdrawn and unable to make
emotional attachments to those around them. Some don't speak or
respond when spoken to, while others simply repeat the words they
hear. Many of these children are thought to be extremely intelligent
rather than retarded, but because they lack the ability to relate to
the world, their intelligence is of little benefit.

Autism used to be blamed on a lack of affection by parents or damage to the brain during delivery, but it's now known that the architecture of the parts of the brain that control memory and emotions are underdeveloped in autistic children. Some scientists have noted that in these children there are reduced numbers of a small but critical group of cells called Purkinje's cells, which are necessary for communication between different areas of the brain.

Researchers at Yale University have shown there is also an over-abundance of the nerve signal substance serotonin in autistic brains. In contrast, serotonin-producing cells are depleted in patients with Alzheimer's disease. In both conditions, information processing is terribly distorted.

To date, little attention has been paid to the possibility that there is a viral cause for autism. But because it appears so arbitrarily in as many as 5 per 10,000 (mostly male) children, a genetic cause has to be ruled out. What we know about the potential effects of viruses on developing fetuses makes it reasonable to think of one as being involved in autism.

Along with the timing of an infection, much of what will happen to a baby depends on the specific agent. Rubella, or German measles, is the classic teratogenic (able to cause abnormal development of a fetus) virus. Mental retardation, cataract, heart disease, deafness, hepatitis, a range of CNS dysfunctions, and failure to thrive have all been caused by the RNA rubella virus. Vaccination against this otherwise minor disease has greatly diminished these awful outcomes in recent years.

The widespread herpesvirus CMV infects as many as 2 of every 100 fetuses and perhaps 4 to 10 babies of every 1000 babies during birth. The timing of the mother's infection—and whether it is a new or a reactivated one—determines the extent of possible damage to the baby. CMV is also acquired during delivery and from breast milk.

Babies severely infected in the uterus by CMV may suffer from neurological problems, deafness, blindness, retardation, spasticity, and various physical deformities. Even moderate CMV infection may leave behind intellectual impairment or subtle behavioral and learning problems that turn up later in childhood.

All the herpesviruses are able to cause CNS damage. Epstein-Barr virus (EBV) infections can result in headache, stiff neck, confusion, reduced consciousness, seizure, and ataxia (movement problems). About 8 percent of people die from EBV infection of the CNS, but the rest usually recover completely.

The herpes simplex 1 virus, which causes genital and oral lesions, sometimes gains access to the CNS, where it routinely produces severe outcomes. Without treatment with the antiviral drug acyclovir, 70 percent die, and even when treated, up to 25 percent succumb to the infection. About a third of those who survive are left with neuropsychiatric problems, many of them serious enough to require institutionalization.

Behavioral problems often resemble those created by an extensive lobotomy (destruction of the temporal lobes of the brain). Amnesia and "striking disorders of behavior and affect" may also follow, leaving patients unable to recognize objects or family members.* Obsessive behavioral and sexual patterns and inappropriate emotional responses are among the many problems caused by changes in the architecture, and therefore the function, of the brain.

When viruses disrupt the normal interplay of chemical messengers between elements of the CNS, although the cognitive function may not be changed appreciably, changes in the nerves that control motor activity result in extreme physical disability, such as that experienced in multiple sclerosis. But there are a surprising number of peculiar CNS disorders, some of which have been found to have a possible association with the newly discovered retroviruses.

Islands of Disease

Degenerative nerve disorders similar to multiple sclerosis (MS) occur in Japan and the tropics. Tropical spastic paraparesis (TSPP) produces a progressive weakness of the lower body and was once thought to be caused by yaws, a spirochetal infection similar to syphilis that is spread by body contact. Significant reduction of yaws cases through World Health Organization efforts, however, has not been followed by a comparable reduction in cases of TSPP.

It was next thought that cassava root, the staple food in many

*E. Kurstak, Z. Lipowski, and P. Morozov, eds., *Virus, Immunity, and Mental Disorders* (New York: Plenum, 1987).

tropical countries, might be causing this degenerative disorder. When not adequately removed by soaking, the cyanide contained in the root produces CNS symptoms. But this theory also failed to explain TSPP when studies showed that those with and without the disorder had identical diets.

Guy de Thé, who tracked Burkitt's lymphoma in East Africa and nasopharangeal cancers in China, came across the association of human T-cell leukemia virus (HTLV) and TSPP by accident while studying patients with leukemia on the island of Martinique. Two of the patients with leukemia examined by de Thé's group were also suffering from TSPP and were positive for the cancer-causing virus. Following this lead, the French scientists examined the blood of seventeen patients who had TSPP but not leukemia. Sixty percent had antibodies against HTLV-1, compared with 4 percent of healthy persons. Taking his investigation to West Africa, de Thé also found tracks of HTLV-1 in TSPP patients there.

Without this serendipitous epidemiological result, no one would have thought to look for a possible association between the cancer-causing virus and the rare neuromuscular diseases. But in Jamaica, Colombia, Trinidad, and the Seychelle Islands off East Africa, where only about 3 percent of the population has antibodies against the leukemia virus, patients diagnosed with TSPP were found to be infected with HTLV. Since HTLV is not widespread in these places—unlike common viruses such as EBV—the fact that patients with TSP are infected indicates that the virus has a causal rather than a casual role.

During research in Jamaica, another unexpected connection was made. Patients with neurological conditions other than TSPP were examined to see whether HTLV was confined to TSPP patients. Six of these control patients were suffering from a condition that has baffled physicians for years—polymyositis, in which muscles are inflamed and painful. All six were HTLV-positive.

Meanwhile, Japanese researchers were looking at the southern-most island of Kyushu, where they found that 15 percent of the people were infected with HTLV and that both adult T-cell leukemia and TSPP were widespread. In addition, Japanese MS patients were found to be HTLV-positive.

Examining the blood of MS patients from Key West, Philadel-phia researcher Hilary Koprowski found that ten of seventeen were

HTLV-positive, along with two of their healthy close contacts. Despite many MS patients having spent time in the Key West hospital and many cases of MS occurring in nurses who had worked there, none of the rest of the staff was positive. A study of Swedish patients with MS and other degenerative neurological diseases caused by damage to the nerve sheaths showed that patients, but not healthy people, had been infected with HTLV.

Is the virus now strongly suspected of causing MS and TSPP the same virus that causes adult T-cell leukemia, or is it a close relative that cross-reacts with the antibody test? The latter possibility seems likely, and recently a virus called HTLV-5 has been found in some patients with degenerative nerve diseases.

Within ten years, then, a whole new category of virus—the retroviruses—has been identified and linked to human cancers, to neuromuscular disorders, and to AIDS. The fact that these viruses remain as silent infections for years and still cause such a diversity of illnesses shouldn't be surprising: We have seen how many common viruses are capable of affecting different body systems in comparably subtle ways. But even these diverse physical outcomes of infection are relatively easy to accept compared with the probability that without any outward signs of what we think of as infection, many viruses cause changes in the way we think.

Mind Viruses

In 1845, a physician at the Saltpetrière in France wrote that "mental alienation is epidemic. There are years when, independent of moral causes, insanity seems suddenly to extend to a great number of individuals."* And after the enormous incidence of severe mental disorders that followed the 1918 influenza epidemic and von Economo's disease, Dr. Karl Menninger, along with many other physicians, was prompted to write that he was "persuaded that dementia praecox [schizophrenia] is at least in most instances a somatopsychosis, the psychic manifestation of an encephalitis."†

At that time the Freudian approach to mental disorders was becoming the rage, and until recent years psychoanalytic thinking dominated research. Emotional and mental problems have been assigned largely to external influences, particularly those experienced during the formative childhood years. While there's little doubt that early experiences determine much of how one deals with life, psychological theories have been unable to explain many of the major psychiatric diseases. Only when new classes of viruses and evidence of their effects on brain cells began to be studied in the

*Quoted in Oliver Sacks, *Awakenings* (New York: Dutton, 1983).
†Ibid.

1970s were the older theories of connections between infectious agents and behavioral disturbances resurrected.

We've seen how acute viral infections can dramatically affect behavior and how damage to the brain's structure may completely change its function and thus cause a variety of behavioral and intellectual disorders. It stands to reason that a substantial number of what are called psychiatric problems may be caused by viruses.

The difficulties of linking a virus to a condition like schizophrenia or manic depression are even more complicated than establishing a cancer-virus connection. Cancers at least produce definitive physical symptoms, and the affected cells can be examined to establish that a cancer is present in specific organs. But how does one even start looking for viruses when deviant behaviors and disturbed thought processes are the only symptoms?

Because disturbances of thinking are filtered through the realities of personality, of different intelligence levels, and often of unique social customs, even an identical underlying cause may manifest and be perceived in totally different ways. What may appear as bizarre behavior in one culture is hardly noticed in another. Going into a trance to act out being an animal is revered in some societies, but the same behavior on a Detroit street corner would bring out the medics with a straitjacket.

In addition to very incomplete knowledge of the brain's chemical milieu, there is an almost complete absence of animal models in which to study subtle brain changes. The difficulties inherent in translating shifts in animal behavior to the behavior of humans have made research agonizingly slow.

In any event, many experts think that the diagnostic labels assigned to most forms of mental dysfunction are merely catchalls for subtler and more diverse kinds of disorders. Depression is a good example. From time to time we all get depressed because of life events that are disappointing or frustrating or beyond our control; how depression is handled depends in part on the personality type or "hardiness" of the individual. But by and large, when the cause is resolved, most people are able to put the depression behind them and get on with their lives.

Other kinds of depression differ from that which accompanies the ordinary ups and downs of life only in duration and intensity. Depression may initially be precipitated by an unhappy event or settle in after an illness such as flu or mononucleosis. When it doesn't lift with the resolution of the problem or illness and leads to subsequent behavioral changes, it becomes a significant and serious disorder, with uncontrollable "manic extravagance and depressive immobility" alternating with each other.*

Manic depression, in which wide swings between elation and despair come and go for no apparent reason, is a distinct entity in the psychiatric literature. It is a condition thought to affect as many as 2 million Americans.

The same symptoms that are present in manic-depressive disease may announce the onset of schizophrenia and psychosis, but they may also presage an infection of the brain. Dissecting out what underlies depression or erratic behavior is therefore difficult, but it is essential if treatment is to be effective. The organic salt lithium is now routinely helping people who suffer from manic depression, and the experience of one psychiatrist illustrates how hard it is to nail down the cause of most of these disorders.

Jay Amsterdam of the University of Pennsylvania recently reported that manic-depressive patients taking lithium experience significantly fewer recurrent genital herpes infections than do patients taking other psychoactive drugs. Since lithium has been shown to suppress viruses in the test tube, a case can be made that in some instances the medicine's effectiveness depends in part on its antiviral actions. However, stress is known to cause the dormant herpesvirus to erupt, so perhaps restoration of emotional balance diminishes the perception of, and response to, some stresses and thus wards off herpes simplex flare-ups. Alternatively, there is abundant evidence of severe emotional problems in certain children who have had a viral infection.

For example, doctors at the Institute of Child Health in London have described four thoroughly examined cases of prepubertal boys who were born with mild developmental disorders, such as slow learning, hyperactivity, and a poor ability to concentrate. After a bout with a viral illness, each of these children developed severe emotional problems. One in particular, a boy named Darren, who was described as being slow but outgoing, active, and affectionate, at age nine had a fever accompanied by headache and confusion.

*Oliver Sacks, *Awakenings* (New York: Dutton, 1983).

At the same time, he also developed a large herpes simplex fever blister.

After recovering from the acute infection, Darren had recurrent episodes of serious manic-depressive mood and behavior disturbances several times a year. They lasted for six weeks at a time and always coincided with outbreaks of fever blisters. Between these episodes, his behavior returned to normal. After several years of this erratic pattern, he was seen by the institute team, which prescribed that the antiviral drug acyclovir be given at the first sign of returning herpes. With the flare-up of herpes short-circuited by the medicine, Darren's behavior remained normal. Proof that the virus was orchestrating his manic-depressive episodes was established when acyclovir was deliberately withheld twice: The fever blister erupted, and Darren at once became disturbed.

Epstein-Barr virus (EBV), another member of the herpes family, is known to cause, and occasionally leave behind, what appear to be psychiatric problems. Depression, intellectual impairment, suicidal thoughts, reduced concentration, feelings of guilt, and loss of fine motor control have all been reported during EBV-caused infectious mononucleosis. Some of these symptoms—depression and anxiety in particular—may hang on for as long as a year after the physical symptoms disappear; infrequently, permanent emotional changes remain.

Chronic Fatigue Syndrome

In late 1984 in the town of Incline Village, Nevada, patients with what seemed to be mononucleosis began appearing in the office of Drs. Paul Cheney and Daniel Peterson. "We weren't particularly impressed at first," Cheney recalls. But they kept coming in, complaining of swollen lymph glands, painful muscles and joints, low-grade fevers, sore throats, headaches, and extreme fatigue. But what frightened many patients was their inability to think.

"These people don't think right: They go to town, and they can't get out because their spatial organization is impaired. They read seven digits in the phone book, and they can't remember long enough to punch them into a touch phone," Cheney explains.

Within nine months the physicians had seen nearly 100 patients with similar symptoms. Most were females, and all were white. There were clusters of cases at several high schools, and in one school the entire girls' basketball team was incapacitated.

Cheney describes two twelve-year-old classmates "who developed acute cerebral ataxia [the staggers] in the same week. They couldn't walk, talk, or speak for several months. They turned out to have demyelinating foci [places where the myelin sheath is eroded] in their brains." Cheney describes the disability accompanying the syndrome as "enormous."

"It wiped out a third of the teachers in one high school and a quarter of another school's faculty, many of whom have still not returned to work in two years," he states.

The *Los Angeles Times* published a story about what was happening in Incline Village, highlighting the memory disturbance and cognitive disorder. At once, Cheney got 150 long-distance telephone calls.

"People said, 'Yes, yes! that's what I've got, and no one believes me! All they do is tell me I'm crazy, that it's all in my mind!' "

From the callers' information, Cheney was able to plot the dates when the illness had begun. A flat line ran from 1953 to 1977, "then in 1977 this thing starts moving in an exponential curve, and by 1984, 1985, 1986, it went right up to the roof," a pattern suggesting a new, fast-moving infectious disease.

Along with the other symptoms, many patients had what appeared to be autoimmune problems. "I think this virus has the capacity to produce autoantibodies against target organs," says Cheney. Physicians are seeing inflammatory heart and thyroid conditions as well as diabetes mellitus caused or pushed along by the illness.

But the amorphous condition that was afflicting the residents of Incline Village and, as it turned out, people all over the country isn't new and isn't well understood. The confusion and frustration of physician and patient alike is succinctly described by James Jones and Bruce Miller of the National Center for Immunology and Respiratory Medicine in Denver in their introduction to a chapter titled "The Postviral Asthenia Syndrome" in *Virus, Immunity and Mental Disorders*.

The authors start with a statement that seems to be the only thing about the condition that can be said with conviction: "The word 'asthenia' is derived from the Greek word *asthenes* [deprived of force] and means loss of energy or strength." A search of the medical literature shows that it has been attributed to "abnormalities in the pituitary gland, muscles, circulatory system, central nervous system, and to Addison's disease, ahydrosis, and constitutional visceroptosis. This wide range of apparent causes for the syndrome tells us immediately that little is really known about the condition, if indeed it is a specific entity."

The multitude of vague symptoms makes it impossible to diagnose patients with a readily recognizable physical illness, and, as the authors put it, "if substantiating physical examination and laboratory findings are absent, definitive labels are difficult or impossible to apply. Patients are frequently informed that they may have had some 'viral illness' and that they will slowly improve." But many do not.

"As these patients continue to frustrate medical practitioners, they are variously labeled as hypochondriacs, malingerers, doctor shoppers or 'cranks,' " the researchers go on to explain.

"Finally, in total frustration, these patients may be advised to seek the advice of a psychiatrist or psychologist with the message that there must be an emotional problem or psychiatric illness as the basis for the symptoms," the authors conclude. In other words, because the patients' symptoms don't match those of a known disease, what they're experiencing ends up being labeled psychoneurotic.

But this same assembly of symptoms was well described almost 100 years ago in Europe, and small outbreaks suggesting an infectious disease pattern have repeatedly attracted attention. Various causes have been proposed, including allergy, metabolic disorders, infectious agents, and psychological problems. Many labels also have been assigned to it: "systemic allergy," "tension-fatigue syndrome," "chronic EBV syndrome," and "epidemic neuromyasthenia."

In 1955, the drawn-out illness of 300 members of the staff of a London hospital was retrospectively blamed on "mass hysteria," which, because the condition affects more women than men, serves as a frequent diagnostic catchall for trivializing what some refer to as female complaints. Several members of the nursing staff remained ill for many years, although they kept silent about their problems after being labeled hysterics.

Describing his patients, Cheney says, "Not a whole lot of them are getting well at two years. I estimate there are maybe 300 or 400 cases with a spectrum of manifestations in our population of 10,000. Take that percentage and put it across the nation and there could be an enormous number of similarly affected people."

Cause or Effect?

While the Nevada doctors were struggling to diagnose their patients, at Northshore University Hospital in Manhasset, New York, Dr. Mark Kaplan diagnosed angioimmunoblastic lymphadenopathy in two patients infected with human immunodeficiency virus (HIV). The precancerous lymph node condition often progresses to the lymphomas seen in people with AIDS. Kaplan suspected that a novel virus might be involved, because when the cells of one of these patients was put in a test tube, they unexpectedly "started growing like crazy." (Only cells infected with EBV had been known to proliferate in test tubes, but that virus wasn't present.)

The second patient had had open heart surgery in 1982 and soon afterward had developed a bizarre rash. Transfusion-related AIDS was suspected, although the man tested negative for HIV antibodies. Kaplan sent both patients' tissues to Zaki Salahuddin at the National Cancer Institute, "who noticed these big, fat cells growing. From that culture he was able to grow out this new virus." As it turned out, the new virus, now called human herpesvirus 6 (HHV-6), represents yet another member of this family of important human pathogens.

Studying blood samples from both healthy and diseased people in many parts of the world, researchers have found that HHV-6 infection is as widespread as is infection with other herpesviruses. To date, HHV-6 has been shown to cause roseola, an acute measleslike disease of infants. The virus may also be implicated as a cofactor in other diseases, such as Hodgkin's lymphoma. To date, however, it has been impossible to connect HHV-6 with other diseases except in a tentative way.

"There is reason to believe that HHV-6 may be an old one in humans and is activated in the course of this illness," says Dr. Anthony Komaroff of Boston, an acknowledged authority on EBV infection whose practice is full of postviral fatigue patients. "We're a long way from establishing that HHV-6 has a causal role," he says. However, in a recent study of 184 patients with what Komaroff describes as "a debilitating illness with neurologic and immunologic abnormalities," brain scans showed that over two-thirds had "punched-out" areas suggesting that demyelination and/or inflammatory processes were taking place.

Komaroff thinks that the HHV-6 infection found in these patients represents the reactivation of an old latent infection. What activates it? No one knows. An unidentified virus may be infecting people who are already carrying latent HHV-6. Alternatively, exposure to the virus after childhood may follow the pattern of late exposure to EBV, another herpesvirus. In Africa, where most very young children are EBV-positive, the high rate of mononucleosis that follows initial EBV exposure of relatively affluent adolescents in developed countries is unknown.

National Cancer Institute epidemiologists say that 22 percent of normal blood donors have antibodies against HHV-6. People with Hodgkin's lymphoma are 79 percent positive, those with acute lymphocytic leukemia are 67 percent positive, those with Burkitt's lymphoma are 87 percent positive, and those with HIV infection are 66 percent positive.

Again, finding traces of a virus is no proof that it is necessarily the cause of a disease; both coxsackie B virus and EBV genetic material are also found in many patients with chronic fatigue syndrome. But there are hints that disordered immune responses are important in causing the symptoms, many of which are like the symptoms experienced by cancer patients being treated with interferon.

The three known interferons, which a few years ago were touted as potential magic bullets against cancers—are produced by immune system cells. These proteins are the body's own antivirals, secreted by infected cells to protect healthy cells in their vicinity. When given systemically, that is, by injection, interferon enters the general circulation and causes a predictable series of symptoms that are almost identical to those of chronic fatigue syndrome. These symptoms include thought and personality changes as well as fatigue, muscle pains, sleep disturbances, and weight loss.

Also, high, continuous levels of interleukin 1 (pyrogen) produced by macrophages in response to persistent viral infection are known to contribute to muscle weakness and wasting. In fact, many aspects of chronic fatigue syndrome suggest that the immune response of the host, in the presence of the virus or viruses that set off the condition, may contribute to the problem. Whatever the precipitating causes, "the ultimate goals of research efforts would be first to benefit the millions of patients worldwide with this syndrome and, secondly, to give us a better understanding of the mechanisms underlying the host's physical and mental responses to infectious diseases," write Jones and Miller.*

Parenthetically, the publication of this book has been delayed by at least a year while I've struggled with what has felt like scrambled brains after a severe bout of flu. The symptoms were too subtle to really nail down: Irritability, gaps in memory, and unaccustomed depression were the prominent ones. One of the physicians to whom I spoke about this section volunteered that the article he was sending me was "six months late."

"I haven't been able to spell or to remember what paragraph went where, since I had flu last winter," he complained.

Fortunately, as these symptoms begin to ease, days in which the brain seems to function as it should outnumber those shadowed by the dullness and disinterest that accompany extreme fatigue. As things return to normal, one can only wonder how people with a comparable long-standing mental deficit can cope. Too many of these patients are dubbed neurotic or psychotic, when the effects or outcomes of viral infection are at the root of their debilitated conditions.

But mental disorders in general seem more open to "interpretation" than those of other body systems. Partly because not much is known about the actual workings of the brain, partly because of changing fashions in what constitutes normal behaviors, and because both patient and physician must struggle with words to describe an ineffable sense of being, the relatively clear-cut descriptions one can apply to problems with the stomach or lungs aren't useful when the sense and seat of self are disrupted.

Research into mental conditions is confronted with the chicken-and-egg conundrum: Which comes first, the changed chemistry that

*J. Jones and B. Miller, in E. Kurstak, et al., eds., *Virus, Immunity and Mental Disorders* (New York: Plenum, 1987).

is often found in mental disorders or changes in function that affect brain chemistry? Are the immune system irregularities found in depression, such as increased levels of cortisone hormones and lower numbers of white blood cells, cause or effect? Schizophrenics have defects in the production of interferons: cause or effect?

The names given to many diseases end in *-itis*, meaning that they are characterized by inflammation. But the process of inflammation is an end point, an outcome, and says little or nothing about causes. There's also nothing particularly judgmental about the suffix *-itis*. But when an illness slides over into the sphere of the mind and brain, other, far less exact elements come into play. Despite progress in biomedical research, little is known about the causes, prevention, and treatment of the many neurological and mental disorders that affect as much as 15 percent of the population. Schizophrenia, the most disabling of these disorders, is a case in point. What probably are numerous dysfunctions are thrown into one basket labeled schizophrenia.

Stalking the Schizovirus

About 1 percent of the world's population suffers from various kinds of mental disorders collectively called schizophrenia. The altered thinking, emotional responses, and behaviors that characterize schizophrenia are most frequently diagnosed in adolescence and early adulthood; when the disorder develops in later life, it usually takes the form of paranoia. It is generally accepted that life events coupled with genetic tendencies are responsible for most cases, but in a fair number of patients there seems to be no clear-cut cause.

A range of symptoms can accompany schizophrenia, from difficulties in thinking logically, to hallucinations, to delusions of persecution or grandeur, to bizarre physical mannerisms and an inability to form relationships. Schizophrenia can mimic either end of the manic-depressive spectrum. Its onset may be accompanied by many of the symptoms associated with viral encephalitis, in which behavioral abnormalities are strikingly like the symptoms of schizophrenia. Until antipsychotic drugs became available, little could be done to help schizophrenic patients, and most treatments are still only marginally effective.

Schizophrenia takes three different courses that are fairly evenly distributed among diagnosed patients. About 25 percent of these people have a single episode from which they recover and thereafter return to normal. Half the remainder (35.5 percent) experience remissions and relapses, while the other half have chronic, progressive disease. Are these all the same disease or are they different ones, and are some of them related to viruses?

You might ask why we should even consider virus as the etiologic agent in this mental disorder when so many family members of schizophrenics tend to have psychiatric illnesses, raising the possibility of a genetic disorder. The answer is that certain aspects of schizophrenia point toward an infective pattern.

The north-south childhood residence differences found in multiple sclerosis (Chapter 15) apply also to schizophrenia in regard to patients living in the northern and southern hemispheres, with their reversed seasons. Among those who develop the disease, a statistically significant number are born in late winter or early spring, suggesting that an agent contracted during the cold months may have an impact on the developing fetus. Alternatively, the immune system of a fetus exposed to a virus early in development may later tolerate the putative virus as part of the self and fail to defend against it.

In addition, many schizophrenics are born with subtle physical differences, such as low-set ears and a single crease in the palm of the hand; about 65 percent are unable to watch a moving object except with a series of jerky eye movements. These minor differences hint at influences on the fetus that may cause various changes in its development. Similar poor eye tracking of moving objects is seen in multiple sclerosis patients, in some patients with brain lesions, in Parkinson's disease patients, and in heavy users of barbiturates and alcohol.

Much of the research involving a possible viral etiology of schizophrenia is still speculative, but intriguing clues have convinced many scientists that if they continue "stalking the schizovirus," as E. Fuller Torrey calls it, there is little question that firm associations will be made in the future.

Torrey, who heads the Twin Research Unit at Saint Elizabeth's Hospital in Washington, D.C., writes that many studies of schizophrenics show various immune system abnormalities that hint at the presence of a viral agent. Combined with the high (10 percent) incidence of serious mental disorders in close relatives of schizophrenics, several factors appear to be at work. One may involve a genetic trait passed down through a family's lineage. Another possibility is that the presence of such a gene interferes with one's ability to handle certain viral infections. Finally, infectious agents actually may be transmitted as parts of genes. Dr. Torrey and National Institutes of Mental Health researchers have recently found that the brains of identical twins—one schizophrenic, the other not— differ; the ill twins have smaller brain volume, particularly in the areas that control the thinking functions.

The recent discovery of slow viruses such as the one that causes AIDS, the apparent attraction of these viruses to central nervous system (CNS) cells, and their habit of becoming persistent and causing diverse diseases in different circumstances (e.g., age and climate), plus advanced technologies that can determine the genetic makeup of these agents, have encouraged us to look at these viruses as potentially important factors in psychiatric disorders.

But the mental disorders that result from many viral infections mimic the entire range of psychiatric diseases. If a biological cause for many of them can be shown, it stands to reason that a biological remedy is in order.

Time Bombs

When HIV was identified, because of its partial similarity to HTLV-1—which a few years earlier had been connected to certain leukemias—the AIDS virus was thought to be a member of that retrovirus family. But researchers soon realized that although no known human virus takes the peculiar course of the one associated with AIDS, the lentiviruses of animal diseases have characteristics similar to those of HIV.

The lentiviruses of sheep and goats are retroviruses that cause slowly progressive diseases involving the CNS, lungs, and joints. These infections are widespread in animal populations throughout the world. The visna-maedi disease of sheep, an inflammatory and demyelinating disease of the CNS, was discovered in Iceland in 1939. Visna virus is structurally related to HIV and shares with it the probable mechanism by which both viruses become persistent.

Caprine arthritis-encephalitis virus (CAEV) was first found in the 1970s in Washington and belongs to the same family as visna. Both infections have a long latency period and result in emaciation, respiratory distress, chronic arthritis, paralysis, or a combination of these symptoms, which almost without exception occur when the animals are adults.

Because slow viruses act just as their name implies, it apparently takes years for the damage to accumulate. Also, as in many other viral infections, when immune system participation is a major factor in a disease, the incompletely developed immune responses of the very young may prevent certain diseases suffered by adults.

Openda Narayan and Janice Clements of Johns Hopkins in Baltimore have shown that the antibodies against visna virus are major factors in the ability of the pathogen to escape destruction by the immune system. Visna plays a bait and switch game in which the virus, under pressure from antibodies, changes its envelope code.

When animals are first infected with visna, they develop acute encephalitis and lesions of the CNS, accompanied by the prompt development of effective antibodies against the virus. However, within months the antibodies that neutralized the parent virus are ineffectual against the mutant strains that subsequently develop. Grown in the test tube without antibodies present, visna virus remains unchanged. But when antibodies are added to the culture, paralleling the situation in infected animals, a series of virus isolates with different coats are produced. Like a compressed time version of what happens to the influenza virus over a period of years, visna reprograms itself to escape destruction by antibodies.

Narayan has shown that when blood from infected sheep is evaluated over time, many distinct new visna types, along with antibodies against them, may appear. This means that the original antibodies fail to recognize the new viruses, which in effect can then cause a new infection. When antibodies develop to handle the new variants, the variants mutate, followed by new antibodies, and so on. With each virus shift and appearance of a novel antibody, more damage is done to the animal; over time, the damage accumulates to produce clinical disease and death.

In experimental animals, "more than forty viruses were recovered from leukocytes [white blood cells] of two sheep during the first ten months after inoculation," Narayan says. The mutations are as capable of causing disease as the parent virus is.

A similar situation obtains in AIDS. Researchers have shown that when blood from a single patient is evaluated month after month, HIV mutates sufficiently to prevent it from being effectively held in check by the original antibodies. Here, as in visna infection, the relentless progress of the disease depends on the interaction between the virus and the response of the immune system to it. Antibody, which should be protective, instead plays a major role in causing the disease process.

Although the animals infected by lentiviruses are different enough to suffer different outcomes of infection, the results are often strikingly similar. All the lentiviruses are able to infect brain cells and to cause demyelination, wasting, arthritis, pneumonias, and CNS disorders. The blood contains few frankly infected cells or little free virus, but circulating immune complexes—amalgamations of virus and antibodies—are usually found. All lentiviruses are transmitted by close contact, and all have long latency periods and the ability to shift their envelope codes to escape immune attack.

Caprine arthritis-encephalitis virus, visna virus, and HIV all infect macrophages in which the viruses can grow for long periods without the cells being killed. But hidden within these phage cells, the virus is protected from the immune system while being transported to target organs.

As researchers begin to close in on the biological bases of mental disturbances in the general, and usually relatively young, population, alarm is also beginning to mount over the incapacitating thought disorders experienced by the growing population of elderly people.

19

Things Fall Apart

The Old Brain

"The most striking indication of the pathology of our species," said science writer Arthur Koestler, "is the contrast between its unique technological achievements and its equally unique incompetence in the conduct of its social affairs."* He equates this with Prometheus reaching for the stars "with an insane grin on his face and a totem-symbol in his hands."

Thus, despite our incredible achievements in biological science, we are stalled at the edge of the age-old chasm, staring in confusion at the ethical and moral dilemmas that underlie philosophy and religion. When does life begin and a fetus become a baby? And when does life end? By what criterion is the decision made to pull the plug on a comatose person? Instead of being argued out in academies of philosophy, these quandaries are now the provenance of politicians seeking reelection, nearly hysterical and polarized interest groups, and scientists whose training ill equips them to adjudicate moral issues.

An equally great and novel quandary is in the making as the

*Arthur Koestler, *The Sleepwalkers* (New York: Macmillan, 1968).

unprecedented life spans of people in the rich western world create a monumental cadre of very old people. In the United States, an estimated 12 million people are sixty-five or older, and those older than eighty-five are the fastest growing sector of the population. What portion of our limited resources can be devoted to their care, and what can be done about the 10 to 15 percent of these people who develop some form of dementia?

Dementia results from progressive degenerative cerebral insufficiency that can be caused by a number of organic factors. In some older people, dementia may consist merely of a slow loss of mental acuity; in others, drastic changes in personality and judgment occur. At the far end of the scale are those whose minds seem simply to have vanished, leaving behind the shell of the person who was.

"Characteristically, orientation becomes impaired first for time, then for place, and finally for person," according to the *Merck Manual*.

Dementia can appear at any age because of meningitis, clogged arteries in the brain, exposure to toxic substances, or severe psychiatric disorders. In most instances, it is a corollary of aging and is based on pathological changes in the brain.

Although distinct changes occur in all elderly brains, such as the "tangles" that are seen in both Alzheimer's disease patients and nondemented persons, relatively few people exhibit a loss of the higher functions, such as memory and language. But both the absolute numbers and the incidence of persons afflicted with these dementing diseases seem to be on the increase. It is the growing *incidence* that adds to the concern: Greater numbers of survivors naturally result in more cases, but greater incidence suggests that the widening spread of external influences must be playing an increasingly prominent role.

A report from Israel reveals that presenile dementia of the Alzheimer type affects 2.4 of 100,000 people aged forty to sixty years. Researcher Theresa Treves says that the incidence is "significantly higher in those born in Europe or America (2.9 percent) than in those born in Africa or Asia." Might early environmental exposures to industrial wastes or a widespread viral agent be implicated, as is suspected in Parkinson's disease?

Parkinson's Disease

The "overheard conversations" described in William Slikker's analogy of dropping a microphone into a building turns into something more like *"Finnegan's Wake* run backward on tape" in degenerative conditions of the aging brain, as Dr. Oliver Sacks described a postencephalitis Parkinson's disease patient.

Alzheimer's-like dementia of the elderly has been noted throughout recorded history, but Parkinson's disease was first described in 1817. Some researchers, including Dr. Sacks, claim that the condition has a 2,000-year history and that Dr. Parkinson's insight merely assembled its many possible symptoms to describe a discrete syndrome.

While physicians agree that the constellation of parkinsonian disorders doubtless has always followed some injuries and viral infections, many insist that Parkinson's disease (PD) itself has appeared relatively recently. Canadian researcher Dr. André Barbeau of Montreal has noted that PD was virtually unknown until the industrial revolution and its toxic wastes appeared in the nineteenth century. Barbeau has studied the contemporary incidence of PD in areas where pesticide use is heavy. He finds a strong correlation between the presence of the disease and the use of agricultural chemicals.

Not only are doctors reporting a greater overall number of PD patients, but many physicians point out that there has been an increase in the past thirty years of as much as 50 percent of PD in younger people.

"It is difficult to avoid the conclusion that there is an environmental risk factor which is becoming more common," insists Dr. Donald Calne of the University of British Columbia.*

Surveys of hospitalized PD patients in Helsinki and Vancouver over a three-year period showed that 8 percent are younger than forty. Half the people with PD suffer from a mild dementia, which in some cases becomes severe. Depression is more commonly present and is often incapacitating.

As well as identifying the aging disease named after Dr. Parkin-

---

*Donald Calne, "Environmental Hypothesis for Brain Diseases Strengthened by New Data," *Science* (July 31, 1987).

son, the term *parkinsonism* is used to describe a set of motor and sometimes behavioral symptoms, much as the term *hepatitis* is used to denote a particular set of liver symptoms as well as the virus that is their primary cause.

Damage to or destruction of a small clump of cells in the brain causes the condition called parkinsonism. This may follow infections or chemical and physical insults or may develop insidiously in aging people. When it is associated with age, the condition is called Parkinson's disease; when it is unconnected to age, the equivalent outcome is referred to as parkinsonism.

Parkinsonism typically includes tremors, muscular rigidity, a blank facial expression, and greasy skin and is particularly noteworthy for the peculiar difficulties it creates when those affected attempt volitional movements such as walking.

Pugilists' Parkinson's disease, which affects former world heavyweight boxing champion Muhammad Ali, often develops in prizefighters and persons who have sustained head injuries, appearing long after the insults to the brain have occurred. Some drugs, such as haloperidol, taken for as little as a few weeks to moderate severe psychiatric symptoms, also can cause parkinsonism, particularly in young males.

Parkinsonism sometimes appears abruptly in tandem with an infection with herpes simplex virus, western equine encephalitis virus, coxsackievirus, or lymphocytic choriomeningitis virus and usually, but not always, resolves over time. But even when they are due to infectious agents, the symptoms of parkinsonism sometimes are delayed for years.

The only recorded pandemic of parkinsonism was von Economo's disease, which followed the 1918 influenza epidemic. However, similar symptoms can follow a number of insults to the area of the brain that regulates the control of the neurotransmitter dopamine, which is essential for normal motor functioning.

Although degeneration of cells in the substantia nigra, an area of large, darkly pigmented cells in the midbrain, was described by von Economo in 1917, damage to this small clump of cells was finally proved to be the underlying cause of PD after a series of drug-related accidents in California in 1982. The story of this discovery began when a forty-two-year-old drug addict was admitted to a hospital in San Jose, California, barely able to speak or move.

A week later his sister, suffering from a less severe though comparable neurological disorder, was also hospitalized. Although both had the classic signs of PD—the expressionless face, slow movements, and tremors—diagnosis was difficult because both were considered too young to have PD.

Neurologists William Langston and Phillip Ballard learned that the siblings had both used a synthetic heroin shortly before their symptoms appeared. However, the doctors were unable to make a firm connection between the drug and the condition of their patients.

But shortly afterward, during a casual conversation with a colleague, Ballard learned that similar symptoms in a young man and his brother were puzzling physicians in another California town. Eventually, what had initially appeared to be a few cases of severe PD in young people turned out to be several dozen. All the patients were drug abusers, and all were permanently, irrevocably disabled.

When the lab-manufactured drug was tested, it was found to contain a chemical called MPTP (1-methyl-4-phenyl-1,2,3-tetrahydropridine), which kills cells of the dopamine-producing substantia nigra. Although the drug accidents had created a neurological disease identical to age-related PD, other dopamine-producing areas of these younger patients' brains showed none of the degenerative changes seen in the naturally occurring form of the disease.

These poisonings provided scientists with absolute proof of the origin of parkinsonism as well as a way to create the disorder in lab animals for the first time: Many animals respond to MPTP just as humans do. When the animal experiments were first carried out, young animals were used to test the effects of MPTP. Although they developed parkinsonism, the particles called Lewy bodies, which are always found in elderly PD patients, weren't seen. When older animals were used, however, they exactly duplicated the disease seen in old humans.

The effects of the synthetic heroin also reinforced studies suggesting that where there is heavy use of industrial chemicals that are related to MPTP (such as the pesticide paraquat), there is also a high incidence of PD.

Research involving identical twins—only one usually has PD— had more or less ruled out a genetic predisposition for contracting the disease. However, studies done after the drug accidents have indicated that there may be an inherited inability to detoxify this class of chemicals, particularly in aging brains.

It's clear that the age at which one develops classic PD says much about its underlying cause. However, vulnerability to developing parkinsonism in response to environmental factors is not necessarily related to age. For instance, in the 1918 encephalitis lethargica pandemic, although the average age was twenty-eight years, many children were affected. When PD is associated primarily with aging, the average patient age is over fifty-five years.

However, in one family, a mother and father and their thirty-seven-year-old son all developed symptoms of PD within a period of three years. The clustering of a degenerative disease in people of different ages strongly suggests an environmental agent which, as is often the case, affects only those with a particular underlying deficit.

The synthetic heroin acted almost immediately, although it resembled age-related PD in its permanent and irreversible changes. The pandemic encephalitis/PD that followed the influenza pandemic—which was certainly caused by a virus, if not the influenza virus, as Ravenholt claims—caused both temporary and permanent disorders.

Those who disagree with Ravenholt's theory argue that the time lag between the two illnesses requires a virus that is able to establish a latent or persistent infection and that the influenza virus clearly doesn't behave that way. But we've seen how different agents can cause PD over widely varying lengths of time.

Viral infection remains among the most compelling culprits in the genesis of PD, although despite numerous investigations, no infective agent has been clearly associated with the disease. Even in the parkinsonism that follows sporadic outbreaks of known viral infections, signs of virus in the brain are usually impossible to find. In age-related PD as well as in Alzheimer's disease, the damage that eventually causes the disease states can take many decades to develop, by which time signs of the agent or substance are long gone.

An environmental agent is known to cause another type of motor neuron disease known as lathyrism, a spastic condition that follows excessive consumption of chickling peas, which have a high content of an amino acid that results in limb weakness. The best known example of this occurred in several hundred former World War II prisoners whose diet consisted primarily of these peas. After more than forty years, many have again begun to exhibit extreme muscle weakness. As in drug-induced cases of parkinsonism, genetics has little to do with what is, frankly, poisoning. In this case, it is age that leads to problems similar to those of people suffering from the postpolio syndrome.

In this country there are approximately 300,000 survivors of polio who contracted the paralyzing disease before the development of a vaccine. Now, twenty and more years later, a quarter of them are experiencing a gradual weakening of their muscles. Some who had been ambulatory are now confined to wheelchairs.

Unlike chickenpox/shingles, the apparent resurgence of polio isn't due to a reactivated virus but rather to an age-related reduction of cells in the anterior horn of the brain. These cells, which in these patients were damaged or depleted years earlier during the acute infection, are now insufficient in number and function to sustain the necessary connections between the brain and motor functions. In effect, although the virus came and went years before, it set in place a situation that is exacerbated by the normal aging process.

Normal aging is associated with accumulated damage to brain neurons. It makes sense, then, that if there has been earlier damage to critical areas of the brain over a lifetime, when cells are routinely lost or no longer function optimally, some effect will be felt. Memory falters, and learning and recall of new information are more difficult for all elderly people.

If actual brain damage has occurred in the past, the natural events accompanying aging may be seriously exacerbated. And if a genetic predisposition for degenerative brain disease exists, depending on the genetic program, any number of conditions may appear. The most feared of these is Alzheimer's disease.

Alzheimer's Disease

In the United States, at least 4 million people have Alzheimer's disease, which robs its victims of their minds and memories. There are two or more kinds of Alzheimer's disease. The first is the familial kind, for which there is known to be a genetic predisposition and which occurs routinely in generation after generation within families. The other type, with identical symptoms, appears sporadically, usually at an earlier age.

Recently, an abnormal gene has been found on chromosome 21

in people with Alzheimer's disease in the family. A single abnormal gene doesn't always mean something, but here it is an important clue, because this gene is close to the location where the abnormal extra gene of Down syndrome lies, and all people with Down syndrome develop Alzheimer's disease if they live past their thirties.

The pathology that causes the disease in Down syndrome and Alzheimer's disease is created by the formation of amyloid plaques and fibrils in the brain, which cause parts of that organ to take on a spongelike appearance. These plaques displace and interrupt the processes by which normal firing of the affected brain cells should take place. The amyloid in Alzheimer's disease and Down syndrome "is composed of a self-aggregating" string of forty-two amino acids.*

According to Dr. Murray Goldstein of the National Institute of Neurological and Communicative Disorders and Stroke, 20 percent of patients with Alzheimer's disease suffered serious head injuries decades earlier. But the most compelling single known risk factor is the presence of a parent or sibling who has the disorder.

Familial Alzheimer's disease accounts for about 10 to 15 percent of cases and can be shown to depend on the inherited gene, as is the case in the familial dementia called Gerstmann-Straussler syndrome (GSS). A study of a family descended from a woman born in 1859 shows that GSS, which is similar to Alzheimer's disease, has never skipped a generation. The degeneration commonly begins in the fourth decade of life—the same time that Huntington's disease appears in those who inherit that genetic trait—and even the brains of family members who have escaped the drawn-out dementia all show some spongiform (spongelike) changes.

Most of these unusual degenerative conditions are sporadic or random, and their genesis is determined by factors other than a genetic predisposition. As far as is known, these degenerative diseases are not transmissible, but given the many years required to develop the conditions, this may not necessarily be true.

In fact, scientists at Yale University have recently reported that they injected hamsters with blood components from eleven relatives of Alzheimer's disease patients, two of whom had "suspicious or

*Carlton Gajdusek, in E. Kurstak, et al., eds., *Virus, Immunity and Mental Disorders* (New York: Plenum, 1987).

early signs of the illness."* The blood from these two, plus three more related persons, caused "biologically documented spongiform encephalopathies in recipient hamsters" identical to the spongiform structures that characterize the human condition.† "These transmission studies raise the intriguing possibility that agents like those in Creutzfeldt-Jakob disease [Chapter 20] may be involved in at least some forms of Alzheimer's disease," the authors conclude.‡

In PD patients there has been some success in at least temporarily controlling symptoms by giving levodopa to replace the loss of activity of the substantia nigra cells, and transplants of dopamine-producing cells hold promise for the future. But Alzheimer's disease affects the broad sheet of cells that constitute the brain's covering cortex, where memories are stored and reasoning powers are generated. The most pronounced damage is done to the cells that produce and use the nerve signal substance acetylcholine.

The brains of older Down syndrome patients as well as those of patients with Alzheimer's disease (AD) have identical pathological characteristics: The orderly neurons, or nerve cells, have been replaced by tangles of twisted, flattened tubular fibrils and by dark bodies called plaques. The same plaques, however, are found in elderly brains in the absence of AD as well as in old monkeys, dogs, and polar bears.

The plaques are composed largely of abnormal proteins called β-amyloid, which is part of a larger protein that serves as a receptor for an unknown substance that interacts with nerve cells. The production of β-amyloid B has also been found to be directed by chromosome 21, and there is hope that eventually the specific defective gene of AD will be found.

On examination of brain sections from AD and Down syndrome patients, it can be seen that the fibrous proteins have taken over the cell body of diseased neurons and apparently are composed of proteins that normally serve as a structural material of the cell walls. The fibrils are also thought to constitute parts of the cell's skeleton.

---

*E. Kurstak, Z. Lipowski, and P. Morozov, *Virus, Immunity and Mental Disorders* (New York: Plenum, 1987).
†Ibid.
‡Ibid.

The greater the number of plaques, the greater the severity of the disease, but what causes these plaques and fibrils to proliferate in one person and not another? Earlier damage to the brain? Toxins? A slow virus?

Clearly, something has gone very wrong in the brains of those who develop this grim loss of self, but no one is sure exactly what. There is a shortage of the hormone somatostatin in affected brains. A protein called A-68 is found in large amounts only in AD brains and can be detected by examining AD patients' spinal fluid. Some think that AD brain cells may lose control over the production of proteins, preventing cellular housecleaning and allowing the buildup of the plaques and fibrils. Others are convinced that with or without genetic susceptibility, AD is caused by environmental agents.

Alternatively, "vulnerability may be determined by the ability of the subjects to detoxify noxious agents or, conversely, to transform a precursor into an active toxin," Canadian researcher Donald Calne wrote in a recent article* (Chapter 17). But the truth of the matter is that despite many recent discoveries of deficient or overabundant hormones and nerve signaling substances in AD patients, little is known about what causes the condition, and nothing at all about how to treat it.

It is well appreciated, however, that the onset of dementia has been the only way to diagnose AD and that by the time this appears, the damage has already been done. Very recently, however, signs of the amyloid protein have been found in the skin and intestines of AD patients, so the time may come when early diagnosis combined with far more knowledge than is currently available will make it possible to at least hold back the degeneration.

But given the influences of the environment and the genes, why even raise the issue of an infectious agent in these degenerative conditions? In the brains of people with AD, PD, and Down syndrome, there are these plaques containing hydroxyapatite–aluminum silicate, a crystalline substance which may signal that this mineral-protein complex is the replicating agent that has proved so elusive.

*Donald Calne, "Environmental Hypothesis for Brain Diseases Strengthened by New Data," *Science* (July 31, 1987).

20

Beyond Virus

Analogies with defective or "contaminated" seed
crystals of simple nucleating molecules specifying
the crystallization of their own distinct crystal
structure come to mind.
> —D. Carlton Gajdusek and C. Joseph Gibbs, Jr.
> in E. Kurstak et al., eds., *Virus, Immunity*
> *and Mental Disorders* (New York: Plenum, 1987)

A "Virus" From the Inorganic World

We're coming full circle, from the viroid of potato plants, through
the world of animal viruses, to the entity described above by its
discoverers, who characterize it as a crystal. They call it a virus
because it's transmissible and can replicate, but it has no genetic
material and is invisible to the immune system. Approaching this
agent—in a gingerly way, because it's barely comprehensible—we
arrive back where we started at the beginning of this book, to the
idea of the spontaneous formation of crystals. But this crystal grows
in our own brains and causes disease if we're very unlucky.

In a funny way, we're back at square one, back 100 years ago,
when the presence of filterable agents was assumed, although none
had yet been seen. The agent described above has been seen but is
not understood. Gajdusek calls it an "unconventional virus," and
Stanley Pruisner of the University of California in San Francisco
chooses the term *prion* for "small proteinaceous infectious particles."
Pruisner maintains that because there apparently is a complete absence of DNA or RNA, it isn't a virus.

Gadjusek vehemently disagrees, stating, "Mathematicians playing with computers have not hesitated to use the term for the 'virus
infection' of computer memories they have produced. The facts
that their software viruses contain no nucleic acid nor are nucleic
acids in any way involved in the pathology that these 'viral diseases'
produce has not prevented computer scientists from appropriately
calling them viruses." But whatever it's called, this agent is responsible for an exotic neurological disease called kuru and other dementias in humans and animals that can be passed from one person
to another.

Ashley Haase of the University of Minnesota describes what, for
the sake of identification (and accommodation between the two
camps), can be called a P-virus as "one of the enigmas of infectious
diseases." The enigma concerns the "increasingly blurred distinction
between infection and pathology."*

If one can conceive of a virus without any genes, one that creates
itself from the cytoskeleton material of cells and causes other proteins to begin accumulating crystalline structures, then it's a virus.
Among its other extreme differences from all known infective
agents, this one is also impervious to the chemicals and radiation
that destroy other viruses, but until a better category is found in
which to place this entity, virus will have to do.

Until I read about this entity, Lyall Watson's description of the
spontaneous crystallization of glycerine seemed more like an intriguing metaphor than a fact, a neat analogy for unknown forces. Not
that crystals don't replicate themselves: Everyone knows that happens, but what does it have to do with viruses? A great deal, if
crystallized glycerine could affect the liquid glycerine in bottles on
the other side of a laboratory, because the crystalline structure of
the P-virus has a similar effect on its uncrystallized counterpart in
the brain.

This exotic agent was discovered in an equally exotic place, the
highlands of New Guinea, by D. Carlton Gajdusek. What Gajdusek
calls a virus belongs to the nonorganic world; it is so peculiar that
it had to be described to an assembly of leading researchers in terms
of tying, or not tying, a knot.

*Ashley Haase, "The Pathogenesis of Slow Virus Infections; Molecular
Analyses," *Journal of Infectious Diseases*, Vol. 153, No. 3 (1987).

"If I loosely tie a square knot in a rope and lay it down, I can shake that knot out. Is it a knot before I pull it tight? But if I pull it tight enough, I'll increase the density fourfold and it will resist even a gasoline fire. I can change all its properties just by increasing its density." This is how Gajdusek explained it.

Bearing an uncanny resemblance to British comedian Benny Hill in a flowered shirt, Gajdusek stormed back and forth across the podium, raising the hair on the scientists' necks. He's got a new paradigm for a new kind of virus with implications far beyond the realm of organic biology.

The idea of viruses being not alive and yet not dead is a barely understandable concept, but when viruses are equated with seeds, it's an idea that is intellectually manageable. But what about a "replicating polypeptide on chromosome 20" (as Pruisner terms it) that copies itself without the benefit of any genetic material?

To help my own understanding of what all this is about, I'm going to describe the degeneration of the brain brought about by the P-virus as a *transmissible process*. It is a process that causes the P-virus to come into being, to crowd out the functioning parts of the brain and nervous system and replace them with inorganic structures.

The P-virus agent is a product of the host's gene (rather than a viral gene) that exists in infected and uninfected animals alike. Gajdusek says the same gene is found "in man and lobsters" and that it hasn't changed "in 100 million years." The gene in question directs the manufacture of amyloid.

Amyloid means "starchlike" and is a substance found in all cells. It comes in two forms. One accumulates in the brains of the victims of Alzheimer's disease and Down syndrome as well as normal elderly people. As we read in Chapter 19, this amyloid is encoded by genes on chromosome 21. The amyloid of the transmissible dementias is specified by genes on chromosome 20.

In Gajdusek's knot analogy, the loose knot represents the routine presence of this second type of amyloid in all healthy older brains. But when the same knot is pulled tight, the plaque accumulates to form a structure with altogether different properties. What pulls the knot? No one is quite sure.

Gajdusek says of the P-virus that whereas in the general population it is likely that 1 per 1 million people a year will develop transmissible dementias such as kuru and Creutzfeldt-Jakob disease, the risk jumps to 1 in 1 "when that configurational change takes place," that is, when the knot is pulled tight. Furthermore, the process of change from a normal to an abnormal amyloid "was put there by iodizing radiation."

If this is the case, the continuing degradation of the ozone layer and the consequent increase in ionizing radiation don't bode well for the growing number of old brains that will be subjected to these events. In other words, an environmental cause, possibly radiation, "twists" the order of amino acids and turns a normal protein into a transmissible agent. The new twist in the rope is in effect "auto-catalytically producing itself in the process," in other words, replicating.

If you take this P-virus and put it in another animal or human, it does the same thing, causing replication of amyloid in the new host. This is known to have happened in the unintentional contamination of corneal transplants and of the growth factor given to children whose physical development is stunted, and in some cases when surgeons operate on elderly brains. In these cases, it is called Creutzfeldt-Jakob disease and it is deadly.

For a while it was thought that because of its small size and resistance to destruction, this minute string of amino acids is related to the "rolling circle" viroid discovered by Theodore Diener. However, as Diener himself reported, this was a "mistaken assumption." In fact, the P-virus differs in many ways from all other transmissible agents.

This agent, all researchers agree, remains infectious in the face of most of the things that can inactivate any known virus, such as ultraviolet radiation, antiviral drugs, high heat, formaldehyde, and nucleases (enzymes that cut amino acids). Because the agent is composed of material from the self rather than utilizing any of its own material, there is no immune response of any kind in infected animals or humans, and no nonhost protein has been found. Even the most unusual virus leaves some of its own material behind in the form of specific viral antigen proteins, but not this one.

Whatever the agent is, it is present in, and probably causes, several rare, progressively dementing diseases in humans. These diseases share many features with transmissible neurodegenerative diseases like scrapie that are known to affect animals. Regardless of the breed of animal in which the agent is found, it's scrapie that creates the disease.

Scrapie

All domestic ruminants—sheep, goats, and cattle—as well as mule deer and elk can contract a fatal disease with an extremely long incubation period. The condition is called spongiform encephalopathy to describe the spongelike degeneration of the brain stem—which controls autonomic activities such as swallowing—and the presence of fibrils in the cerebral cortex, where many thought processes arise. These animal diseases are characterized by gradual changes in nervous system signs and chronic weight loss.

Scrapie has been recognized as an animal disease for more than 200 years, but only in the 1980s has it appeared in cows. In England, 9,000 cases of bovine spongiform encephalopathy (BSE) have been found in the herds and at least 600 new cases are appearing each month.

There are only a few cases of scrapie in any herd of animals, although transmission is known to occur between sheep grazing in contaminated pastures. There is some suspicion that feeding the cows bonemeal and other products from infected animals might have caused the disease to spread to them, but there is no proof of this. However, scrapie is a transmissible agent that has been given to experimental animals, including mice and hamsters.

Regardless of what the disease is called or which animal is injected with the scrapie agent, the infective material is coded for by the animal, not by the agent. This is precisely what occurs when the P-virus agent acts within human brains.

The scrapie agents likewise are rod-shaped and crystalline and are almost identical to the amyloid plaques and fibrils found in Alzheimer's disease, Creutzfeldt-Jakob disease (CJD), chronic wasting disease of mule deer and elk, and kuru. It was this last disease—kuru—that alerted the scientific community to the presence of an infective agent which, transmitted from person to person, can cause a dramatic neurodegenerative disease in humans.

Kuru in the New Guinea Highlands

In 1976, D. Carlton Gajdusek shared the Nobel prize in physiology and medicine with Baruch Blumberg (Chapter 10). Blumberg had discovered the first definitive tracks of one of the world's most widespread viral diseases—hepatitis B—and Gajdusek discovered what probably is the rarest disease of all among the stone-age tribal people in New Guinea.

The population of Fore-speaking natives, of which there was a total of 35,000, suffered from what "we originally called juvenile galloping senesence, because these kids of ten or eleven had massive amyloid plaques," Gajdusek recalls.*

Kuru in the Fore language means "trembling." This was the first sign that the deadly condition had commenced: Within a year, routinely and without fail, death followed. First the trembling, then an inability to control the limbs, then into the terminal stage. The researchers soon saw that children and women of all ages, but few men, were afflicted by kuru.

Gajdusek observed that ritualistic cannibalism was practiced as part of the mourning ceremony for a dead relative. It was carried out primarily by the women and participated in by the children, and highly infectious material from the brains was easily transmitted by skin contamination.

This was one instance in which a primitive people benefited from moving into the twentieth century. Between the cessation of the ritual funerary practices in 1957 and 1975, more than 2,500 died of kuru. But gradually, as the Fore have turned to coffee farming, the incidence of kuru has lessened dramatically.

Kuru disappeared in young children five to nine years old during the first five years of change. Then the population aged ten to fourteen years ceased having kuru, then the adolescents. No child born since the end of ritual cannibalism has been affected, and cases in adults have gradually dropped to a negligible number. The dementing disease has ceased to exist because the source of infection is no longer passed among the people.

Theresa Elizan and Jordi Casals of Mount Sinai Hospital in New York have written that after the discovery of kuru, "it became pos-

---

*D. Carlton Gajdusek in E. Kurstak et al., eds., *Virus, Immunity and Mental Disorders* (New York: Plenum, 1987).

sible to conceive that other progressive subacute and chronic neurological diseases of man, previously regarded as basically degenerative and noninflammatory, could be potentially infectious in
etiology, possibly virally induced, and that a prolonged latent interval, measured in years, between initial experience with a virus
and such neurological disorders may not be particularly unusual."*
In other words, some of the dementias that have been accepted as
being "the way it goes" in the elderly may instead be the long-term
outcome of an atypical infective agent.

Despite intensive investigations, no one was able to find any trace
of what was thought to be the slow virus that caused these deadly
brain disorders, which were clearly passed from one person to another. Now the science community has to grapple with the possibility
that "this mineral-protein complex is the replicating agent that has
proved so elusive," insists Pruisner.

Even Gajdusek admits that "we'll never be able to tell you what
this is: It'll be crystallographers measuring bond angles and tertiary
and secondary configurations of crystallized proteins" who explain
this newest infectious viruslike agent, an entity that weaves itself
into what Gajdusek calls the "macramé on a symphony of molecular
large yarn."† This extraordinary agent of change in days to come
will, it is hoped, make sense of the "wastebasket" of degenerative
brain disease.

The Future

Knowledge about this self-replicating array of crystals surely will
bring into closer association that which we think of as "living" and
"biological" and those entities we have always considered inert and
nonliving. If viruses themselves are not alive yet not dead, what of
crystals that cause and transmit disease? Given the utter strangeness
of the P-virus, what does tomorrow hold? A bit of that already
glimmers out at the far edge of science.

They've injected a mouse with a virus, taken the antibody the

---

*Theresa Elizan and Jordi Casals, in E. Kurstak et al., *Virus, Immunity and
Mental Disorders* (New York: Plenum, 1987).
†D. Carlton Gajdusek, in E. Kurstak et al.

mouse produces against the virus, cloned the antibody's genes, and installed one part in one plant and one part in another plant via a special bacterium with an affinity for greenery. When the two plants are mature, they are cross-pollinated and their offspring churn out antibodies, thus accomplishing what no plant has ever been able to do before.

Since plants have no immune systems to protect themselves against the pathogens that prey on them, once this technology is applied, the plants will be able to neutralize pollutants and resist damaging insects. In the long term, it's likely that antibodies that target human cancer cells will be grown in Dr. Diener's potato plants.

If we can take advantage of what now seems like peace breaking out, if we unclench and turn our attention away from antagonism and toward accommodation, if we use the technology and ideas that have awaited funding, even the word *virus* may someday soon have a beneficent rather than a frightening connotation. Instead of stockpiling these dangerous pathogens for some doomsday assault on our fellows, with a little luck and a lot of work, we will understand viruses in such a way that their unique qualities will be utilized for the good of this planet and its inhabitants.

Glossary

Acquired immune deficiency syndrome (AIDS): An infectious disease characterized by failure of the immune system caused by HIV.

Allergen: Any substance that causes an allergy.

Allergy: An inappropriate immune system response to harmless substances.

Alzheimer's disease: A loss of mind and memory that affects the aged.

Anaphylaxis: A violent allergic reaction rarely seen on repeat exposure to an antigen.

Antibiotics: Drugs that assist the body in killing bacteria, some fungi, protozoa, and other microorganisms.

Antibody: Protein molecules that are produced by B cells (B lymphocytes) in response to an antigen.

Antigen: Any substance that provokes an immune response.

Arbovirus: A family of viruses transmitted by arthropods (insects).

Arthritis: A range of specific diseases or symptoms of other diseases that cause inflammation of the joints and/or muscles.

Attenuated: Used to describe a microbe that has been changed or weakened so that it does not cause disease.

Autoantibody: An antibody that reacts against a person's own tissues.

Autoimmune disease: A disease caused by an immune system attack on the self.

Bacterium: A single-celled organism capable of causing disease.

B cell: A white blood cell that produces antibodies to create immunity against certain infectious agents.

Bone marrow: The soft tissue found in the hollow center of long bones in which blood cells are produced.

Brain tumor: An abnormal growth in the brain that is sometimes cancerous.

Burkitt's lymphoma: A cancer of the lymphatic tissue associated with Epstein-Barr virus. Mainly affects young males in equatorial Africa.

Cancer: Abnormal growth of tissues that can affect any part of the body.

Candidiasis: Infection with a yeastlike fungus called *Candida*.

Capsid: The protein covering that covers the nucleic acid core of a virus.

Case-control study: An epidemiological method in which persons with a disease condition are compared with a healthy population similar in age, sex, race, etc., to determine the differences between them.

Cell: The smallest independent unit of an organism. A cell is composed of cytoplasm and a nucleus and is surrounded by a membrane or wall.

Cell-mediated immunity: Immunity conferred by T lymphocytes.

Centrifuge: A rapidly spinning lab machine that separates elements from one another.

Chemotherapy: Treatment of a disease with chemicals, such as antibiotics and cytotoxic drugs.

Chromosome: A structure containing genetic material (DNA) within the nucleus of a cell.

Circulating immune complexes: Conglomerates of antigen and antibody found in the blood.

Clone: A group of cells derived from a single parent cell and identical to it.

Clotting factors: Substances in the blood responsible for normal coagulation.

Complement: A series of proteins in the blood that participate in the immune response, particularly the inflammatory response.

Congenital: Said of a condition present at birth; the opposite of acquired.

Corticosteroids: Natural substances that work as hormones and are used as pharmacological agents to treat many conditions with an inflammatory component.

Cryptococcus: A fungus that seldom causes infection in healthy persons.

Cytomegalovirus (CMV): A widespread herpesvirus capable of producing severe disease in infants and persons with a suppressed immune function.

Cytoplasm: The watery material between the nucelus and the membrane of a cell.

Cytotoxic drugs: Chemical substances with a toxic or killing effect on cells.

Dementia: A progressive failure of cognitive function caused by organic factors.

DNA: The double helix of genetic material that carries genetic information.

Dopamine: A substance that functions as a transmitter of signals between CNS cells.

Ebola: An infectious virus found in parts of equatorial Africa. Related to the agent that causes Marburg disease.

Encephalitis: Inflammation of the brain; usually caused by an infectious agent.

Endemic: A condition or disease that is widespread in a population.

Epidemic: An outbreak of disease among a population.

Epitope: A characteristic shape or marker on the surface of an antigen.

Epstein-Barr virus (EBV): A herpesvirus that is responsible for infectious mononucleosis and is thought also to play a role in Burkitt's lymphoma and nasopharyngeal cancers.

Factor VIII: One of the clotting factors in the blood. Congenital absence of factor VIII causes hemophilia A.

Gamma globulin: One of the proteins in the blood serum that contains antibodies.

Gastroenteritis: Inflammation of the lining of the stomach and intestines.

Genetic marker: Any substance whose presence is an indicator of an inherited disorder.

Gland: An organ that produces specialized chemicals, such as hormones, that are released into the blood to act at distant sites.

Globulins: Simple proteins found in the blood serum which contain various molecules central to immune system function.

Graft-versus-host reaction (GVH reaction): The immune system attack against grafted tissues.

Helper T cells: Lymphocytes bearing the CD4 marker that are responsible for many immune system functions, including turning antibody production on and off.

Hemophilia: A hereditary disease in which one or more of the clotting factors are missing from the blood.

Hepatitis: Various infections of the liver caused by one of several viruses, e.g., hepatitis viruses A, B, and C and the delta fraction.

Herd immunity: Passive innoculation against an infectious disease passed from immunized members of a group to unimmunized members of the group.

Herpes: A family of large viruses that includes herpes simplex, herpes zoster, EBV, CMV, and human herpes virus-6 (HHV-6). All can cause disease in humans.

Histocompatibility antigens (HLA): A group of molecules found on the membranes of cells that are specific to the individual. They are used to determine compatibility of tissues in organ transplants and participate in the immune response to viruses.

HLA (Human Leukocyte Antigen): Genetic markers on cells.

Hodgkin's lymphoma: A cancer of the lymph system.

Host defense: The sum of bodily protection against disease.

Human immunodeficiency virus (HIV): The slow virus that causes AIDS.

Human T-cell leukemia virus (HTLV): A recently identified family of slow viruses known to cause several conditions, including cancers, in humans.

Humoral immunity: The resistance to disease conferred by the antibodies produced by B lymphocytes.

Immune complexes: Combinations of antibodies and antigens that circulate in the blood: found in certain infectious and autoimmune diseases.

Immune system: A complex of cells and proteins that helps fight infectious diseases and regulates many of the body's functions.

Immunization: Protection against disease by vaccination, usually with a weakened form of a pathogen that is unable to cause illness.

Immunoglobulins: Serum proteins that confer immunity.

Immunosuppressive drugs: Chemicals that lower the immune response and are given to prevent organ transplant rejection and treat overactive responses that cause autoimmune diseases.

Incubation time: The period between exposure to an infectious agent and the first signs of infection.

Interferons: Chemical substances produced by cells in response to infection; they also serve as local antiviral agents. Interferons produced by cloning are used to treat some cancers.

Interleukin-1 (IL-1) A molecule that is activated early in an immune response: also called progen because it causes fever.

Interleukin-2 (IL-2): One of a family of molecules that control the growth and function of many types of lymphocytes.

Isolate: The particular strain of an infectious agent; used to describe the specific agent found in an individual.

Kaposi's sarcoma (KS): A cancerlike growth of small blood vessels; may be lethargic in older people or aggressive in young Africans and patients with AIDS.

Killer T cells: Lymphocytes able to kill cells that are foreign, cancerous, or infected.

Kuru: A rare degenerative infectious disorder of the brain and central nervous system caused by an uncharacteristic virus, or prion.

Latency: The period between contracting a disease and showing symptoms; similar to incubation period.

Legionnaires' disease: An acute infectious pneumonia caused by a water-borne bacterium.

Lesion: Any abnormality of a tissue or organ; may be a visible or metabolic abnormality.

Leukemia: Cancer of the white blood cells and bone marrow.

Leukocytes: White blood cells.

Lymph nodes: Small organs of the immune system that are widely distributed throughout the body.

Lymphocytes: Small white blood cells that bear major responsibility for carrying out immune system functions.

Lymphoma: Cancer of the lymphoid tissues.

Macrophage: A scavenger cell found in the tissues that cooperates with other white cells in the immune response.

Mast cells: Special cells that produce the symptoms of allergy.

Memory cells: T cells that have been exposed to specific antigens and are able thereafter to proliferate upon repeat exposure to the same antigens.

Meningitis: Inflammation of the membranes that cover the brain and spinal cord.

Metabolism: The chemical processes that take place in the body.

Microbes: Microscopic organisms: bacteria, viruses, etc.

Mitogen: A substance that induces cell growth and division.

Monoclonal antibodies: Antibodies produced in the laboratory that are specific for a single antigen.

Mononucleosis: The "kissing disease" caused by the Epstein-Barr virus.

Mutation: A change in genetic information, either spontaneous or produced by external influences such as radiation.

Mycobacterium: A bacterium that frequently is the cause of pneumonia.

Myeloma: A cancer of antibody-producing cells in the bone marrow.

Natural killer cells: Large lymphocytes that attack and destroy infected and tumorous cells.

Neoplasm: A tissue that is proliferating in an uncontrolled fashion; a cancer or malignant tumor.

Oncogene: A gene that is able to cause cancer.

Opportunistic infection: An infection in immune-suppressed people caused by a common microorganism.

Pandemic: An epidemic that has spread widely.

Parasite: A plant or animal that lives and feeds on or within another living organism.

Parkinson's disease: A chronic degenerative or temporary condition characterized by loss of control of movement.

Parvovirus: A family of small viruses known to cause disease in animals and people.

Passive immunity: Transfer of immune competent antibody-contain ing substances from one person to another.

Peyer's patches: A collection of lymphoid cells in the intestinal tract.

Plasma: The fluid part of the blood; contains minerals and proteins.

Pneumocystis carinii pneumonia: A type of pneumonia seen frequently in AIDS patients; caused by fungus.

Polyclonal: Having an origin in more than one type of ancestor cell. Contrast with Monoclonal.

Protein: Organic compounds composed of the amino acids from which life is created.

Protozoan: A single-cell parasite.

Retrovirus: A virus with RNA as its genetic material; requires the enzyme reverse transcriptase to reproduce itself.

RNA: A form of genetic material complementary to DNA.

Scavenger cells: Any of a diverse group of cells with the capacity to engulf and destroy foreign material and dead tissues and cells.

Schizophrenia: A severe mental condition in which the thinking process is disorganized.

Scrapie: A neurodegenerative disease of animals caused by an agent that is unlike a virus (also see Kuru).

Severe combined immune deficiency disease (SCIDS): A disease of infants who are born with varying degrees of incompetent immune systems.

Spectrum: In infectious diseases, the range of severity of infection, from symptomless to severe.

Stem cells: The bone marrow cells from which all blood cells are produced.

Suppressor T cells: A subset of T cells that down-regulate the immune response.

T cells: The family of lymphocytes that are responsible for fighting viral and fungal infections.

Thymus: A small organ in which T cells are programmed for their functions. The site of production of thymic hormones.

Titer: The greatest dilution of antigen or antibody that produces the desired result in a blood test.

Toxoplasmosis: An infectious disease produced by a protozoan that seldom causes disease in healthy people.

Vaccine: A suspension of weakened or killed bacteria or virus that is given to protect against later exposure to the same pathogen.

Vector: An animal or insect that carries and passes a disease-causing organism.

Viremia: The presence in the blood of viruses.

Viroid: A small infectious agent of certain plants: reproduces by "rolling" copies of itself.

Virus: A subcellular organism composed of genetic material and protein; able to reproduce only within living cells.

References

This book draws on the research of scientists in many disciplines. Much of the work described here has been published in medical and science journals and in collections of reports from conferences, a number of which I have attended. Among the researchers who have generously spent time answering my endless questions are the following:

Jay D. Amsterdam, Depression Research Unit, Department of Psychiatry, School of Medicine, University of Pennsylvania, Philadelphia, Pennsylvania

Baruch Blumberg, Fox Chase Cancer Center, Philadelphia, Pennsylvania

Arsesne Burny, Laboratoire de Chimie Biologique, Université Libre, Brussels, Belgium

Thomas Caskey, Laboratory of Molecular Genetics, Baylor University, Houston, Texas

Paul Cheney, private practice, Charlotte, North Carolina

Theodore Diener, Plant Protection Institute, U.S. Department of Agriculture, Beltsville, Maryland

Robert Gallo, Laboratory of Tumor Cell Biology, National Cancer Institute, Bethesda, Maryland

John Gerin, Division of Molecular Virology and Immunology, Georgetown University, Rockville, Maryland

Allan Goldstein, Department of Biochemistry, George Washington University School of Medicine and Health Sciences, Washington, D.C.

Ludwik Gross, Cancer Research Unit, VA Medical Center, Bronx, New York

George Hensley, Department of Pathology, Jackson Memorial Hospital, Miami, Florida

Martin Hirsch, Department of Medicine, Massachusetts General Hospital, Boston, Massachusetts

Terry Johnson, Professor and Director, Division of Biology, Kansas State University, Manhattan, Kansas

Mark Kaplan, Northshore Hospital, Manhasset, New York

David Katz, Medical Biology Institute, La Jolla, California

Anthony Komaroff, Brigham and Women's Hospital, Boston, Massachusetts

Hilary Koprowski, Wistar Institute of Anatomy and Biology, Philadelphia, Pennsylvania

Hiroaki Mitsuya, National Institute of Diabetes, Digestive, and Kidney Diseases, Bethesda, Maryland

Luc Montagnier, Institut Pasteur, Paris, France

Openda Narayan, Department of Pathology and Neurology, Johns Hopkins University School of Medicine, Baltimore, Maryland

Michael Oldstone, Department of Immunology, Scripps Clinic and Research Foundation, La Jolla, California

R. A. Ravenholt, National Institute on Drug Abuse, Rockville, Maryland

Zaki Salahuddin, Laboratory of Tumor Cell Biology, National Cancer Institute, Bethesda, Maryland

Petr Skrabanek, School of Public Health, Dublin, Ireland

William Slikker, National Center for Toxicological Research, Little Rock, Arkansas

Raphael B. Stricker, Department of Medicine, University of California School of Medicine, San Francisco, California

Guy de Thé, Laboratoire d'Epidemiologie des Tumeurs, Lyons, France

E. Fuller Torry, Neuropsychiatry Branch, National Institute of Mental Health, Saint Elizabeth's Hospital, Washington, D.C.

Dimitri Trichopoulous, Harvard University School of Medicine, Boston, Massachusetts

Bibliography

Barrett, J., ed., *Textbook of Immunology*, 5th ed. St. Louis: Mosby, 1988.

Blumberg, B. S., "Les Prix Nobel en 1976." Stockholm, Sweden: The Nobel Foundation, 1977.

Goldstein, A., ed., *Thymic Hormones and Lymphokines*. New York: Plenum, 1984.

Hartwig, G., and Patterson, D., *Disease in African History*. Durham, N.C.: Duke University Press, 1978.

Koestler, A., *The Sleepwalkers*. New York: Macmillan, 1968.

Kurstak, E., Lipowski, Z., and Morovov, P., eds., *Virus, Immunity and Mental Disorders*. New York: Plenum, 1987.

Lawler, G., and Fischer, T., eds., *Manual of Allergy and Immunology*. Boston: Little Brown, 1981.

McNeill, W., *Plaques and People*. Garden City, N.Y.: Anchor, 1976.

Merck Manual, 13th ed. Rahway, N.J.: Merck & Co., 1978.

Notkins, A., and Oldstone, M., eds., *Concepts in Viral Pathogenesis*. New York: Springer-Verlag, 1984.

Rizzetto, M., Gerin, J., and Purcell, R., eds., *The Hepatitis Delta Virus and Its Infection*. New York: Alan Liss, 1987.

Sacks, O., *Awakenings*. New York: Dutton, 1983.

Stringfellow, D., ed., *Virology*. Kalamazoo, Mich.: Scope, 1983.

Watson, L., *Lifetide*. New York: Bantam, 1980.

Williams, G., *Virus Hunters*. New York: Knopf, 1960.

Wolfgram, F., Ellison, G., Stevens J., and Andrews, J., eds. *Multiple Sclerosis*. New York: Academic Press, 1972.

Youmans, G., Paterson, P., and Sommers, H., eds., *The Biological and Clinical Basis of Infectious Diseases*, 2nd ed. Philadelphia: Saunders, 1980.

Index

Accessory cells, 59–60
Acetylcholine, 192, 251
Acute infections, 29
Acute lymphocytic leukemia, 236
Acyclovir, 226, 232
Adenosine deaminase, 110; deficiency, 74
Adenoviruses, 29, 80, 129, 154
Adsorption, 27
Adult T-cell leukemia (ATL), 103, 203, 227–228
Aedes aegypti, 205–206
A. taeniorhynchus, 204
African swine fever virus (ASFV), 182, 207
Aging process, 222; causes of, 217; normal, 249
Agricultural Research Center, 36
Agriculture, U.S. Department of (USDA), 36, 38, 182
AIDS (acquired immune deficiency syndrome), 1, 2, 28, 32, 48–49; autoimmune diseases and, 61–64, 199–200; and brain, 220–221, 222; cause of, 5, 8; and

AIDS (acquired immune deficiency syndrome) (*Cont.*):
Centers for Disease Control, 128; different effects of, 72; early attitudes toward, 3, 181–183; first appearance of, 3, 4–5, 86; and human herpes, 6, 30, 68; impact on science of, 8–9; infighting over, among research groups, 85–86; "legitimizing" of, 22; papers published about, 84–85; and parvovirus, 183–184; as reflection of social and technological changes, 4; and retroviruses, 228; spread of, 7; and viruses, 240; and white blood cells, 54. *See also* Human immunodeficiency virus (HIV)
Alar (daminozide), 71
Ali, Muhammad, 246
Allergen, 190–191
Allergies, 52, 190–191
ALS (amyotrophic lateral sclerosis), *see* Lou Gehrig's disease
Altman, Lawrence, 86